Introduction to Partial Differential Equations

Donald Greenspan
University of Texas at Arlington

DOVER PUBLICATIONS, INC.
Mineola, New York

Bibliographical Note

This Dover edition, first published in 2000, is an unabridged reprint, with minor corrections, of the first edition of *Introduction to Partial Differential Equations,* published by McGraw-Hill Book Company, New York, in 1961.

Library of Congress Cataloging-in-Publication Data

Greenspan, Donald.
    Introduction to partial differential equations / Donald Greenspan. — Dover ed.
       p. cm.
    Originally published: New York : McGraw-Hill, c1961.
    Includes bibliographical references and index.
    ISBN 0-486-41450-7 (pbk.)
       1. Differential equations, Partial. I. Title.

QA374 .G714 2000
515'.353—dc21

                                  00-031573

Manufactured in the United States of America
Dover Publications, Inc., 31 East 2nd Street, Mineola, N.Y. 11501

PREFACE

This book is designed for use in a one-semester course in partial differential equations for seniors and beginning graduate students in the mathematical, physical, and engineering sciences. It requires for its study only a facility with those concepts of real-valued functions usually explored in advanced calculus, such as one-sided limit, greatest lower bound, uniform continuity, multiple integral, and absolute convergence. The presentation is rigorous and explores both practical methods of solution and the unifying theory underlying the mathematical superstructure. Heuristic appeal is made, wherever possible, to geometric intuition even though the equations considered have extensive physical application. None of the material is new but some of it is relatively modern. Much of the classical content has been adapted, synthesized, and extended from the excellent works of Churchill [5], Courant and Hilbert [9], Horn [22], John [25], Lax [29], Petrovsky [37], and Tamarkin and Feller [47].

Inevitably, the aims set forth above dictate the deletion or deemphasis of certain important topics. For example, the adoption of the succinct subharmonic-superharmonic function approach of Perron for the study of the Dirichlet problem relegated a discussion of the Green's function to the final, cursory chapter, while the extensive functional analysis prerequisites necessary for a study of the elliptic-parabolic theory of Fichera [12] relegated this ultramodern ingress to the limbo of the bibliography. The reader is encouraged therefore to explore the various source materials listed after Chap. 8.

Concerning the teaching of the text, note that Chaps. 1 and 2 merely develop those aspects of ordinary differential equations, complex variables, and Fourier series which are essential for the study of partial differential equations, which begins properly in Chap. 3. One may therefore delete those sections of Chaps. 1 and 2 with which the student is already familiar.   Also, certain discussions, notably those in Secs. 2.1 and 5.4, are somewhat more sophisticated than others, and the instructor may wish to show discretion concerning their presentation.

Finally, I wish to thank J. B. Diaz, P. M. Anselone, and, in particular, R. M. Warten, for their excellent constructive criticisms of the original manuscript.

Donald Greenspan

CONTENTS

Introduction to
Partial Differential
Equations

Chapter 1

BASIC CONCEPTS

1.1 Sets

One of the fundamental concepts of modern mathematical thought is that of a set. From the strictly logical point of view, the term *set* may be left undefined. Intuitively, however, one may think of a set as a collection, or aggregate, of objects called elements or points.

Example 1. The positive integers 1, 2, 3, 4, . . . constitute a set.

Example 2. All polynomials with rational coefficients constitute a set.

Example 3. All real numbers which satisfy the inequality $0 \le x \le \pi$ constitute a set.

Example 4. All points (x,y) of the plane whose coordinates satisfy the inequality $x^2 + y^2 \le 1$ constitute a set.

The set of no elements is called the empty set and is denoted by θ.

Example. The set of all real numbers x which satisfy $x^2 < 0$ is θ.

Definition 1.1. The union, or sum, of two sets A and B, denoted $A \cup B$, is the set of all elements which belong to at least one of A and B.

Example 1. If A is the set whose elements are the integers 1, 2, 3 and B is the set whose elements are the integers 3, 4, 5, then $A \cup B$ is the set whose elements are the integers 1, 2, 3, 4, 5.

Example 2. If A is the set of rational numbers and B is the set of irrational numbers, then $A \cup B$ is the set of real numbers.

Definition 1.2. The union, or sum, of an arbitrary collection of sets A_α, denoted $\cup_\alpha A_\alpha$, is the set of all elements which belong to at least one of the A_α.

1

Definition 1.3. The intersection, or product, of two sets A and B, denoted $A \cap B$, is the set of all elements which belong to both A and B.

Example 1. If A is the set whose elements are the integers 1, 2, 3 and B is the set whose elements are the integers 3, 4, 5, then $A \cap B$ is the set which consists of the single integer 3.

Example 2. If A is the set of rational numbers and B is the set of irrational numbers, then $A \cap B = \theta$.

Example 3. If A is the set of all points (x,y) of the plane whose co-ordinates satisfy the inequality $x^2 + y^2 \leq 1$ and B is the set of all points (x,y) of the plane whose coordinates satisfy the inequality $x^2 + y^2 \leq 2$, then $A \cap B$ is the set of all points (x,y) of the plane whose coordinates satisfy the inequality $x^2 + y^2 \leq 1$.

Definition 1.4. The intersection of an arbitrary number of sets A_α, denoted $\underset{\alpha}{\cap} A_\alpha$, is the set of all elements which belong to every A_α.

Definition 1.5. Two sets A and B are said to be equal if and only if they contain exactly the same elements.

Definition 1.6. If every element x of a set A is also an element of a set B, then one says that A is contained in B, or B contains A, or A is a subset of B, and one writes $A \subset B$.

Example. If A is the set of all points (x,y) of the plane whose co-ordinates satisfy the inequality $x^2 + y^2 \leq 1$ and B is the set of all points (x,y) of the plane whose coordinates satisfy the inequality $x^2 + y^2 \leq 2$, then $A \subset B$.

Definition 1.7. If $A \subset B$ and B contains an element b which is not an element of A, then A is said to be a proper subset of B.

Example 1. If A is the set of rational numbers and B is the set of real numbers, then A is a proper subset of B.

Example 2. If A is the set of real numbers and B is the set of real numbers, then A *is* a subset of B but is *not* a proper subset of B.

Theorem 1.1. $A = B$ if and only if both $A \subset B$ and $B \subset A$.

Proof. If $A = B$, then A and B contain exactly the same elements, so that $A \subset B$ and $B \subset A$. Conversely, if $A \subset B$ and $B \subset A$, then every element of A is an element of B and every element of B is an element of A, so that A and B contain the same elements. Hence, $A = B$, and the theorem is proved.

It will be convenient as the discussion develops to use the notation "$x \in A$" to mean "x is an element of the set A."

1.2   Plane Point Sets

For completeness, important types of point sets in the plane will now be reviewed. The notions to be considered, however, are so basic that they can be extended readily to point sets in three, four, or more dimensions.

Throughout, the plane will be denoted by the symbol E^2.

Definition 1.8. In E^2, a circular neighborhood of a fixed point P with coordinates $(\bar{x}, \bar{y})$ is the set of all points (x, y) which, for some fixed, positive r, satisfy the relationship $(x - \bar{x})^2 + (y - \bar{y})^2 < r^2$.

A convenient notation at present for a circular neighborhood will be $K(P, r)$.

Example. If P has coordinates $(1, -1)$, then $K(P, 3)$ is the set of all points (x, y) which satisfy the relationship $(x - 1)^2 + (y + 1)^2 < 9$.

Note that, in Definition 1.8, r is always finite and positive and that $(\bar{x}, \bar{y})$ is itself an element of $K(P, r)$.

Definition 1.9. In E^2, a point P with coordinates $(\bar{x}, \bar{y})$ is said to be an interior point of a set M if and only if there exists at least one circular neighborhood $K(P, r)$ such that $K(P, r) \subset M$.

Example. Let M be the set of all points (x, y) whose coordinates satisfy the inequality $x^2 + y^2 \leq 1$. Then $(0, 0)$, $(\frac{1}{2}, 0)$, $(0, -\frac{3}{4})$, $(\frac{1}{4}, \frac{7}{8})$ are all interior points of M, while $(1, 0)$, $(1, 1)$, $(-1, 3)$ are not interior points of M.

Definition 1.10. In E^2, a set of points M is said to be open if and only if each of its points is an interior point of M.

Example 1. The set of points (x, y) whose coordinates satisfy $x^2 + y^2 < 1$ is an open set.

Example 2. The set of points (x, y) whose coordinates satisfy $x^2 + 2y^2 < 1$ is an open set.

Example 3. The set of points (x, y) whose coordinates satisfy $x^2 + y^2 > 1$ is an open set.

Example 4. The set of points (x, y) whose coordinates satisfy $x^2 + y^2 \neq 0$ is an open set.

Example 5. The set of points (x, y) whose coordinates satisfy $0 < x < \pi$, $0 < y$ is an open set.

Example 6. E^2 is itself an open set.

Example 7. θ is an open set.

Definition 1.11. In E^2, the complement of a set of points M, denoted by $E^2 - M$, is the set of all points of E^2 which are *not* elements of M.

Example. If M is the set of points (x,y) whose coordinates satisfy $x^2 + y^2 < 1$, then $E^2 - M$ is the set of points (x,y) whose coordinates satisfy $x^2 + y^2 \geq 1$.

Definition 1.12. In E^2, a set of points M is said to be closed if and only if $E^2 - M$ is open.

Example 1. The set of points (x,y) whose coordinates satisfy $x^2 + y^2 \leq 1$ is a closed set.

Example 2. The set of points (x,y) whose coordinates satisfy $1 \leq x^2 + y^2 \leq 2$ is a closed set.

Example 3. The set of points (x,y) whose coordinates satisfy $x \geq 0$, $y \geq 0$ is a closed set.

Example 4. E^2 is a closed set.

Example 5. θ is a closed set.

Note that in common usage the terms *open* and *closed* are antonyms, whereas this is *not* true in mathematical usage. For example, E^2 is *both* open and closed. Note further that the set M which consists of all points (x,y) whose coordinates satisfy $0 < x < \pi$, $0 \leq y \leq \pi$ is *neither* open *nor* closed.

Definition 1.13. In E^2, a point P is said to be a limit point of a set M if and only if every circular neighborhood of P contains at least one point of M which is different from P.

Example. Let M be the set of all points (x,y) whose coordinates satisfy $x^2 + y^2 < 1$. Then $(0,0)$, $(\frac{1}{2},0)$, and $(1,0)$ are limit points of M, while $(2,0)$, $(-3,-3)$ are not limit points of M.

Note in Definition 1.13 that a point P can be a limit point of a set M and yet not be an element of the set M.

Theorem 1.2. In E^2, a necessary and sufficient condition that a nonempty set M be closed is that it contain all its limit points.

Proof. Suppose M is closed. Assume there exists a point P which is a limit point of M but which is an element of $E^2 - M$. By definition, $E^2 - M$ is open and P is therefore an interior point of $E^2 - M$. Hence, for some $\epsilon > 0$, there exists $K(P,\epsilon) \subset (E^2 - M)$. But no point of M lies in $K(P,\epsilon)$, so that P could not have been a limit point of M. By contradiction, then, M contains all its limit points.

Suppose next that M contains all its limit points. In order to prove that M is closed, it is sufficient to prove that $E^2 - M$ contains only interior points. Suppose then there is a point P of $E^2 - M$ which is not an interior point of $E^2 - M$. Then every circular neighborhood $K(P,\epsilon)$ contains a point P_α of M. The points P_α

and P can never be the same since $P \in (E^2 - M)$, $P_\alpha \in M$, and $M \cap (E^2 - M) = \theta$. Hence, P is a limit point of M, which is a contradiction. Hence, $E^2 - M$ contains only interior points and M is closed, which completes the proof.

Theorem 1.3. In E^2, if M_1, M_2 are nonempty closed sets, then $M = M_1 \cup M_2$ is closed.

Proof. Let P be a limit point of M. Then for every $\epsilon = 1/n$, $n = 1, 2, 3, \ldots$, there exists a point P_n, different from P, such that $P_n \in [M \cap K(P,\epsilon)]$. The sequence $\{P_n\}$ is contained in $M_1 \cup M_2$ and contains an infinite number of distinct points. Hence, at least one of M_1 or M_2, say M_1, contains an infinite number of distinct points of the sequence. By Definition 1.3 and Theorem 1.2, it follows that $P \in M_1 \subseteq M$, so that M is closed. Since an analogous argument follows for the choice of M_2 in place of M_1, the theorem follows readily.

Definition 1.14. In E^2, a set of points M is said to be bounded if and only if there exists a fixed point P and a positive real number r such that $M \subseteq K(P,r)$.

Example 1. Let M be the set of all points (x,y) whose coordinates satisfy the relationships $0 < x < \pi$, $0 \le y \le \pi$. Then M is bounded.

Example 2. Let M_1 be the set of points (x,y) whose coordinates satisfy $x^2 + y^2 < 1$, and let M_2 be the set of points (x,y) whose coordinates satisfy $(x - 1)^2 + 2(y - 1)^2 < 1$. Then $M = M_1 \cup M_2$ is a bounded set.

Theorem 1.4. In E^2, if M_1, M_2 are two bounded sets, then $M = M_1 \cup M_2$ is a bounded set.

Proof. The proof follows directly from the fact that in E^2 a circular neighborhood always exists which contains any two given circular neighborhoods.

Definition 1.15. In E^2, a boundary point P of a set M is a point for which every circular neighborhood $K(P,r)$ contains at least one point of M and at least one point of $E^2 - M$.

Example 1. If M is the set of all points (x,y) whose coordinates satisfy $x^2 + y^2 < 1$, then the points $(1,0)$ and $\left(\dfrac{1}{\sqrt{2}}, \dfrac{1}{\sqrt{2}} \right)$ are boundary points of M, while $(0,0)$, $(\tfrac{1}{2},\tfrac{1}{2})$, $(-2,4)$ are not boundary points of M.

Example 2. If M is the set of all points (x,y) whose coordinates satisfy $0 < x^2 + y^2 \le 1$, then $(1,0)$, $\left(\dfrac{1}{\sqrt{2}}, \dfrac{1}{\sqrt{2}} \right)$, $(0,0)$ are boundary points of M, while $(\tfrac{1}{2},\tfrac{1}{2})$, $(-2,4)$ are not boundary points of M.

Definition 1.16. In E^2, the union of all boundary points of a set M is defined to be the boundary of M.

Example 1. If M is the set of all points (x,y) whose coordinates satisfy $x^2 + y^2 < 1$, then the boundary of M is the set of all points (x,y) whose coordinates satisfy $x^2 + y^2 = 1$.

Example 2. If M is the set of all points (x,y) whose coordinates satisfy $x^2 + y^2 \leq 1$, then the boundary of M is the set of all points (x,y) whose coordinates satisfy $x^2 + y^2 = 1$.

The boundaries of the plane point sets to be considered in connection with partial differential equations will be called *contours*. In order to make the notion of a *contour* precise, we consider first a special class of plane curves called *bounded* plane curves.

Definition 1.17. Let a, b be two real numbers, with $a < b$. Let $x = f(t)$, $y = g(t)$ be two real-valued, continuous functions of real parameter t, where $a \leq t \leq b$ and where at least one of $f(t)$, $g(t)$ is nonconstant. Then, in E^2, the set of points (x,y) whose coordinates are determined by the relationships

$$x = f(t), \quad y = g(t), \qquad a \leq t \leq b, \tag{1.1}$$

is said to define a *bounded plane curve*, or, more precisely, a *continuous bounded plane curve*, and Eqs. (1.1) are called parametric equations of, or a parametric representation of, the curve.

Example 1. $x = \sin t, y = \cos t, 0 \leq t \leq \pi$ is a parametric representation of a bounded plane curve called a semicircle.

Example 2. $x = \sin t, y = \cos t, 0 \leq t \leq 2\pi$ are parametric equations of a bounded plane curve called the unit circle, that is, of the circle with center $(0,0)$ and radius equal to one.

Note that a bounded plane curve may have a variety of parametric representations.

Example. $x = \cos t, y = \sin t, \qquad 0 \leq t \leq \pi;$

$$x = \sin t, y = \cos t, \qquad -\frac{\pi}{2} \leq t \leq \frac{\pi}{2};$$

$$x = t, y = \sqrt{1 - t^2}, \qquad -1 \leq t \leq 1$$

are three parametric representations of the same semicircle.

Also note that the condition $a < b$ and the condition that at least one of $f(t)$, $g(t)$ be nonconstant imply that any bounded plane curve must consist of more than one point.

Definition 1.18. For a bounded plane curve, given by (1.1), the two points $(f(a), g(a))$, $(f(b), g(b))$ are called the end points of the curve.

Example. (1,0) and $(-1,0)$ are the end points of the semicircle defined by $x = \cos t, y = \sin t, 0 \le t \le \pi$.

Definition 1.19. A point $(\bar{x}, \bar{y})$ of a bounded plane curve given by (1.1) is said to be a double point if and only if there exist at least two *distinct* parameter values t_1, t_2, such that $a \le t_1 \le b, a \le t_2 \le b$ and $\bar{x} = f(t_1) = f(t_2), \bar{y} = g(t_1) = g(t_2)$.

Intuitively, a double point is a point where a curve "intersects," "crosses," or "meets" itself.

Example. $x = \left(\frac{\sqrt{2}}{2} + \cos t \right) \cos t, y = \left(\frac{\sqrt{2}}{2} + \cos t \right) \sin t, 0 \le t \le 2\pi$ are parametric equations of a limaçon, or, perhaps more descriptively, of a cardioid with a loop (consult Diagram 1.1). Parametric values $t_1 = 3\pi/4$, $t_2 = 5\pi/4$ yield the double point $(0,0)$, while parametric values $t_1 = 0$, $t_2 = 2\pi$ yield a second double point $\left(1 + \frac{\sqrt{2}}{2}, 0 \right)$.

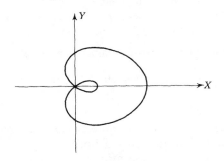

DIAGRAM 1.1

Definition 1.20. A bounded plane curve is said to be closed if and only if its end points coincide.

Example. The limaçon given by $x = \left(\frac{\sqrt{2}}{2} + \cos t \right) \cos t, y = \left(\frac{\sqrt{2}}{2} + \cos t \right) \sin t, 0 \le t \le 2\pi$ is closed.

Definition 1.21. A bounded plane curve is said to be simple if and only if no points, except possibly its end points, are double points of the curve.

Example 1. The semicircle given parametrically by $x = \cos t, y = \sin t$, $0 \leq t \leq \pi$ is simple.

Example 2. The circle given parametrically by $x = \cos t$, $y = \sin t$, $0 \leq t \leq 2\pi$ is simple.

Definition 1.22. A bounded plane curve is said to be a simple, closed curve if and only if it is both simple and closed.

Example 1. The circle with parametric representation $x = \cos t, y = \sin t$, $0 \leq t \leq 2\pi$ is a simple, closed curve.

Example 2. The ellipse with parametric representation $x = 2 \cos t$, $y = 3 \sin t$, $0 \leq t \leq 2\pi$ is a simple, closed curve.

Example 3. Let $f(t)$, $g(t)$ be the continuous functions defined on $0 \leq t \leq 6$ by

$$f(t) = \begin{cases} t, & 0 \leq t \leq 1 \\ 1, & 1 \leq t \leq 3 \\ 4 - t, & 3 \leq t \leq 4 \\ 0, & 4 \leq t \leq 6 \end{cases} \qquad g(t) = \begin{cases} 0, & 0 \leq t \leq 1 \\ t - 1, & 1 \leq t \leq 3 \\ 2, & 3 \leq t \leq 4 \\ 6 - t, & 4 \leq t \leq 6. \end{cases}$$

Then $x = f(t), y = g(t), 0 \leq t \leq 6$ are parametric equations of a well-known type of simple, closed curve called a rectangle.

Example 4. The semicircle with parametric representation $x = \cos t$, $y = \sin t$, $0 \leq t \leq \pi$ is *not* a simple, closed curve.

Example 5. The limaçon with parametric representation $x = \left(\dfrac{\sqrt{2}}{2} + \cos t\right) \cos t$, $y = \left(\dfrac{\sqrt{2}}{2} + \cos t\right) \sin t, 0 \leq t \leq 2\pi$ is *not* a simple, closed curve.

Definition 1.23. A bounded plane curve given by (1.1) is said to be *sectionally smooth* if and only if both

    (*a*) relative to at least one continuous parametric representation of form (1.1), for each, except possibly a finite number of, t in the range $a \leq t \leq b, f(t)$ and $g(t)$ have continuous first- and second-order derivatives and, at the exceptional points, the corresponding right- and left-hand derivatives exist;

    (*b*) whenever each exists, neither $[(df/dt)^2 + (dg/dt)^2]$ nor the corresponding sum of right- or left-hand derivatives is zero.

Example 1. The straight-line segment given parametrically by $x = 1$, $y = t, 0 \leq t \leq 2$ is sectionally smooth.

Example 2. The circle given parametrically by $x = \cos t$, $y = \sin t$, $0 \leq t \leq 2\pi$ is sectionally smooth.

Example 3. Let $f(t)$, $g(t)$ be the continuous functions defined on $0 \leq t \leq 4$ by

$$f(t) = \begin{cases} t, & 0 \leq t \leq 1 \\ 1, & 1 \leq t \leq 3 \\ 4-t, & 3 \leq t \leq 4 \end{cases} \qquad g(t) = \begin{cases} 0, & 0 \leq t \leq 1 \\ t-1, & 1 \leq t \leq 3 \\ 2, & 3 \leq t \leq 4. \end{cases}$$

Then $x = f(t)$, $y = g(t)$, $0 \leq t \leq 4$ are parametric equations of a sectionally smooth curve.

Example 4. Let $f(t)$, $g(t)$ be the continuous functions defined on $0 \leq t \leq 4$ by

$$f(t) = \begin{cases} t, & 0 \leq t \leq 2 \\ 2\cos\left(\dfrac{t-2}{2}\right)\pi, & 2 \leq t \leq 3 \\ 0, & 3 \leq t \leq 4 \end{cases}$$

$$g(t) = \begin{cases} 0, & 0 \leq t \leq 2 \\ \sin\left(\dfrac{t-2}{2}\right)\pi, & 2 \leq t \leq 3 \\ 4-t, & 3 \leq t \leq 4. \end{cases}$$

Then $x = f(t)$, $y = g(t)$, $0 \leq t \leq 4$ are parametric equations of a sectionally smooth curve.

Note that, in Definition 1.23, assumption (b) is related to the existence of *nonzero* tangent vectors, while the assumption of the existence and continuity of the second-order derivatives is related to "continuous turning" of a tangent vector. Note also that Definition 1.23 is of such generality as to include as a special case the usual concept of a *smooth* curve.

Definition 1.24. In E^2, a simple, closed curve which is sectionally smooth is called a *contour*.

Example 1. The circle given parametrically by $x = \cos t$, $y = \sin t$, $0 \leq t \leq 2\pi$ is a contour.

Example 2. Let $f(t)$, $g(t)$ be the continuous functions defined on $0 \leq t \leq 6$ as follows.

$$f(t) = \begin{cases} t, & 0 \leq t \leq 1 \\ 1, & 1 \leq t \leq 3 \\ 4-t, & 3 \leq t \leq 4 \\ 0, & 4 \leq t \leq 6 \end{cases} \qquad g(t) = \begin{cases} 0, & 0 \leq t \leq 1 \\ t-1, & 1 \leq t \leq 3 \\ 2, & 3 \leq t \leq 4 \\ 6-t, & 4 \leq t \leq 6. \end{cases}$$

Then $x = f(t)$, $y = g(t)$, $0 \leq t \leq 6$ are parametric equations of a contour.

Example 3. Let $f(t)$, $g(t)$ be the continuous functions on $0 \leq t \leq 4$ which are defined as follows.

$$f(t) = \begin{cases} t, & 0 \leq t \leq 2 \\ 2 \cos \left(\dfrac{t-2}{2}\right)\pi, & 2 \leq t \leq 3 \\ 0, & 3 \leq t \leq 4 \end{cases}$$

$$g(t) = \begin{cases} 0, & 0 \leq t \leq 2 \\ \sin \left(\dfrac{t-2}{2}\right)\pi, & 2 \leq t \leq 3 \\ 4-t, & 3 \leq t \leq 4. \end{cases}$$

Then $x = f(t), y = g(t), 0 \leq t \leq 4$ are parametric equations of a contour.

With the aid of the concept of a bounded curve, one can also define precisely, in addition to *contour*, the useful concept of *connectedness*.

Definition 1.25. In E^2, a set M is said to be connected if and only if for every pair of points P_1, P_2 of M there exists at least one continuous bounded curve C such that P_1, P_2 are end points of C and $C \subset M$.

Example 1. The set of points (x,y) whose coordinates satisfy $x^2 + y^2 < 1$ is a connected set.

Example 2. The set of points (x,y) whose coordinates satisfy $1 \leq x^2 + y^2 \leq 4$ is a connected set.

Example 3. The set of points (x,y) whose coordinates satisfy $-\pi \leq x \leq \pi, y \leq 0$ is a connected set.

Example 4. Let M_1 be the set of points (x,y) whose coordinates satisfy $x^2 + y^2 < 1$ and let M_2 be the set of points (x,y) whose coordinates satisfy $(x-4)^2 + (y-4)^2 < 1$. Then $M = M_1 \cup M_2$ is *not* connected.

The following fundamental theorem, called the Jordan curve theorem, relates the concept of connectedness with that of a simple, closed curve.

Theorem 1.5.   Jordan Curve Theorem. In E^2, let C be a simple, closed curve. Then $E^2 - C$ consists of exactly two open, connected sets A, B, only one of which is bounded, such that $A \cap B = \theta$ and such that C is the boundary of both A and B.

The proof of this fundamental theorem, despite all intuitive evidence, is quite deep and cannot be given here. Complete details are available in Hall and Spencer [18].

Note also that of the two sets A, B described in Theorem 1.5, the bounded one is usually called the inside or interior of C while the other is usually called the outside or exterior of C.

Example 1. Let the unit circle C be given parametrically by $x = \cos t$, $y = \sin t$, $0 \leq t \leq 2\pi$. Let A be the set of all points (x,y) whose coordinates satisfy $x^2 + y^2 < 1$ and let B be the set of all points (x,y) whose coordinates satisfy $x^2 + y^2 > 1$. Then $E^2 - C = A \cup B$; A and B are both open and connected; only A is bounded; $A \cap B = \theta$, and C is the boundary of A and of B. Since A is bounded and B is not, A is the interior of C while B is the exterior of C.

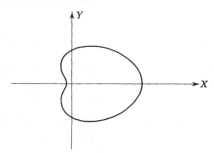

DIAGRAM 1.2

Example 2. Consider the cardioid C with parametric representation $x = (2 + \cos t) \cos t$, $y = (2 + \cos t) \sin t$, $0 \leq t \leq 2\pi$ (consult Diagram 1.2). For fixed $t = \bar{t}$, consider the set of points (x,y) defined by $x = v(2 + \cos \bar{t}) \cos \bar{t}$, $y = v(2 + \cos \bar{t}) \sin \bar{t}$, $0 \leq v < 1$. Let A be the set of all such points determined for all $\bar{t}$ in $0 \leq \bar{t} < 2\pi$. For fixed $t = t^*$, consider the set of points (x,y) defined by $x = u(2 + \cos t^*) \cos t^*$, $y = u(2 + \cos t^*) \sin t^*$, $u > 1$. Let B be the set of all such points determined for all t^* in the range $0 \leq t^* < 2\pi$. Then $E^2 - C = A \cup B$; A and B are both open and connected; only A is bounded; $A \cap B = \theta$, and C is the boundary of A and of B. Since A is bounded and B is not, A is the interior of C while B is the exterior.

With the aid of the Jordan curve theorem, a valuable classification of connected sets can now be established. Consider, for example, the following two sets. Let M_1 be the set of all points (x,y) whose coordinates satisfy $x^2 + y^2 \leq 1$ and let M_2 be the set of all points (x,y) whose coordinates satisfy $0.01 \leq x^2 + y^2 \leq 1$ (consult Diagrams 1.3 and 1.4). Both M_1 and M_2 are connected sets, and yet there is a basic difference between the two, for, intuitively speaking, M_2 has a hole in it while M_1 does not. To differentiate between these possibilities, the following definition is given.

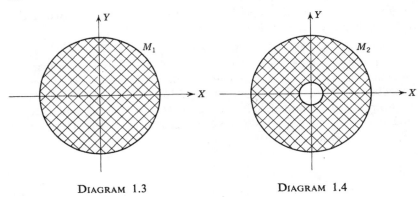

DIAGRAM 1.3   DIAGRAM 1.4

Definition 1.26. In E^2, a connected set M is said to be *simply connected* if and only if for every simple, closed curve C contained in M, the interior of C is also contained in M.

Example 1. Let M be the set of all points (x,y) whose coordinates satisfy $x^2 + y^2 \leq 1$. Then M is simply connected.

Example 2. The interior of every simple, closed curve is a simply connected set.

Example 3. Let M be the set of all points (x,y) whose coordinates satisfy $0.01 \leq x^2 + y^2 \leq 1$. Then M is connected, but is *not* simply connected.

Example 4. The exterior of every simple, closed curve is connected, but is *not* simply connected.

Another type of plane point set which has varied uses is the *region*.

Definition 1.27. In E^2, a *region* is any nonempty, connected, open set.

Example 1. The set R of all points (x,y) whose coordinates satisfy $x^2 + y^2 < 1$ is a region.

Example 2. The set R of all points (x,y) whose coordinates satisfy $1 < x^2 + y^2 < 2$ is a region.

Example 3. The set R of all points (x,y) whose coordinates satisfy $x > 0$, $y > 0$ is a region.

Example 4. E^2 is itself a region.

Definition 1.28. The union of a region R and its boundary B is called a *closed region* and is denoted by $\bar{R} = R \cup B$.

Example. If R is the set of all points (x,y) whose coordinates satisfy $x^2 + y^2 < 1$, then R is a region and $\bar{R}$ is the set of all points (x,y) whose coordinates satisfy $x^2 + y^2 \leq 1$.

Note that a closed region is a closed set.

Finally, after some preliminary considerations, a major theorem about plane point sets, the Heine-Borel theorem, will be proved.

Definition 1.29. In E^2, let $\{\bar{R}_n\}$ be a sequence of square, closed regions. Then if, for all n, $\bar{R}_{n+1} \subset \bar{R}_n$, one says the sequence $\{\bar{R}_n\}$ is a *sequence of nested closed squares*.

Example. The sequence $\{\bar{R}_n\}$ of square, closed regions constructed by letting the vertices of $\bar{R}_n$ be $(0,0)$, $\left(\dfrac{1}{n}, 0\right)$, $\left(0, \dfrac{1}{n}\right)$, $\left(\dfrac{1}{n}, \dfrac{1}{n}\right)$ is a sequence of nested closed squares.

Lemma 1.1. In E^2, let $\{\bar{R}_n\}$ be a sequence of square, closed regions such that each $\bar{R}_n$ has sides which are parallel to the coordinate axes. Let d_n be the length of the diagonal of $\bar{R}_n$. If $\bar{R}_{n+1} \subset \bar{R}_n$, for each n, and if $\lim\limits_{n \to \infty} d_n = 0$, then there exists a unique point $(\bar{x}, \bar{y})$ such that $\bigcap\limits_{n=1}^{\infty} \bar{R}_n = (\bar{x}, \bar{y})$.

Proof. The projection of each $\bar{R}_n$ on the x axis is a closed interval. Let the left-hand end point of this interval be x_n and let the right-hand end point be x_n'. Since $\{\bar{R}_n\}$ is a nested sequence, $\{x_n\}$ is an increasing sequence of real numbers and $\{x_n'\}$ is a decreasing sequence of real numbers. Since $\{x_n\}$ is bounded above and $\{x_n'\}$ is bounded below, $\{x_n\}$ and $\{x_n'\}$ are convergent. Since $\lim\limits_{n \to \infty} d_n = 0$, $\{x_n\}$ and $\{x_n'\}$ converge to the same value, say $\bar{x}$. In a similar fashion, projection of each $\bar{R}_n$ on the y axis yields a corresponding value $\bar{y}$, and it readily follows that the point $(\bar{x}, \bar{y})$ is the unique point in question. The lemma is thus proved.

Theorem 1.6. Heine-Borel Theorem. In E^2, let M be a non-empty, closed, bounded point set. Let there be given a collection of *open* sets S_α such that $M \subset \bigcup\limits_{\alpha} S_\alpha$. Then there exists a *finite number* of sets $S_{\alpha_1}, S_{\alpha_2}, \ldots, S_{\alpha_k}$ among the sets S_α such that $M \subset \bigcup\limits_{n=1}^{k} S_{\alpha_n}$.

Proof. If M consists of a finite number of points, then the theorem follows readily, for one need only choose one open set of S_α for each point of M in order to construct the desired finite number of open sets.

Suppose M consists, then, of an infinite number of points and assume that no finite number of open sets selected from the S_α exists which will satisfy the theorem.

Since M is bounded, there exists a square, closed region $\bar{R}_1^{(1)}$, with sides parallel to the coordinate axes, such that $M \subset \bar{R}_1^{(1)}$. Let $(x_1, y_1) \in [M \cap \bar{R}_1^{(1)}]$.

Quadrisect $\bar{R}_1^{(1)}$ into four congruent, square, closed regions $\bar{R}_2^{(1)}$, $\bar{R}_2^{(2)}$, $\bar{R}_2^{(3)}$, $\bar{R}_2^{(4)}$. Consider $\bar{R}_2^{(1)} \cap M$, $\bar{R}_2^{(2)} \cap M$, $\bar{R}_2^{(3)} \cap M$, $\bar{R}_2^{(4)} \cap M$. At least one of these four sets, say $\bar{R}_2^{(1)} \cap M$, is not contained in the union of a finite number of the S_α and contains an infinite number of points of M. Let $(x_2, y_2) \in [\bar{R}_2^{(1)} \cap M]$ and let (x_2, y_2) be distinct from (x_1, y_1). Next quadrisect $\bar{R}_2^{(1)}$ into four congruent, square, closed regions $\bar{R}_3^{(1)}$, $\bar{R}_3^{(2)}$, $R_3^{(3)}$, $\bar{R}_3^{(4)}$. Consider $\bar{R}_3^{(1)} \cap M$, $\bar{R}_3^{(2)} \cap M$, $\bar{R}_3^{(3)} \cap M$, $\bar{R}_3^{(4)} \cap M$. At least one of these four sets, say $\bar{R}_3^{(1)} \cap M$, is not contained in the union of a finite number of the S_α and contains an infinite number of points of M. Let $(x_3, y_3) \in [\bar{R}_3^{(1)} \cap M]$ and let (x_3, y_3) be distinct from (x_1, y_1), (x_2, y_2). In an inductive fashion, then, construct, as indicated above,

(a) a sequence $\{\bar{R}_n^{(1)}\}$ of nested closed squares,

(b) a sequence of sets $\{\bar{R}_n^{(1)} \cap M\}$ such that no element of the sequence is contained in the union of a finite number of the S_α and each of which contains an infinite number of points of M,

(c) a sequence $\{(x_n, y_n)\}$ of distinct points such that $(x_n, y_n) \in [\bar{R}_n^{(1)} \cap M]$, $n = 1, 2, 3 \ldots$.

The sequence $\{\bar{R}_n^{(1)}\}$ is a sequence of nested closed squares each of which has sides parallel to the coordinate axes. Moreover, if d_n is the length of the diameter of $\bar{R}_n^{(1)}$, then by construction $\lim\limits_{n \to \infty} d_n = 0$.

By Lemma 1.1, there exists unique point $(\bar{x}, \bar{y})$ such that $\overset{\infty}{\underset{n=1}{\cap}} \bar{R}_n^{(1)} = (\bar{x}, \bar{y})$. Moreover, by construction, $(\bar{x}, \bar{y})$ is a limit point of the sequence $\{(x_n, y_n)\}$, so that, since M is closed, $(\bar{x}, \bar{y}) \in M$.

Hence, there exists at least one open set S^* amongst the S_α such that $(\bar{x}, \bar{y}) \in S^*$. Since S^* is open, $(\bar{x}, \bar{y})$ is an interior point of S^*. There exists, therefore, a circular neighborhood K^* of $(\bar{x}, \bar{y})$, of radius $r^* > 0$, such that $(\bar{x}, \bar{y}) \in K^* \subset S^*$. There exists, however, positive integer N such that for all $n > N$, $R_n^{(1)}$ has length of diameter less than $\frac{1}{2} r^*$. Hence, for all $n > N$, $(\bar{x}, \bar{y}) \in [\bar{R}_n^{(1)} \cap M] \subset \bar{R}_n^{(1)} \subset K^* \subset S^*$. This contradicts the fact that each set $[\bar{R}_n^{(1)} \cap M]$ could not be covered by a finite number of the S_α, and the theorem is proved.

1.3 Remarks on Real Functions

The concepts of function, limit, continuity, derivative, and integral for functions of two variables are now recalled. It is assumed that the reader has been exposed to the basic theorems germane to these concepts and that he will recognize the validity of the mathematical structure when such theorems are applied to the study of partial differential equations.

Definition 1.30. If to every ordered pair (x,y), where x is an element of a nonempty set M_1 and y is an element of a nonempty set M_2, there corresponds by some rule a unique element u of a set M_3, then a function f, of two variables, is said to be defined. The element u of M_3 which corresponds under the rule to the ordered pair (x,y) is called the value of f at (x,y), and one writes $u = f(x,y)$.

Although the concepts of *function* and of *value of a function* are quite distinct, these two ideas are usually identified in the mathematical literature so that we shall denote functions by any one of: f, $f(x,y)$, u, or $u(x,y)$.

If the sets M_1, M_2, M_3 in Definition 1.30 are subsets, proper or improper, of the set of real numbers, then $u = f(x,y)$ is said to be a *real*, or *real-valued*, function, and, unless otherwise stated, the term *function* in the sequel will imply *real-valued function*.

Example 1. $u = x^2 + y^2$ defines a real function for all real x, y.

Example 2. $u = x + \sqrt{y}$ defines a real function for all real x and for all real nonnegative y.

Example 3. For all real x, y such that $x^2 + y^2 \leq 1$, $u = \sqrt{1 - x^2 - y^2}$ defines a real function.

Geometrically, of course, the ordered pair (x,y) of Definition 1.30 may be considered as a point in the xy plane. The graph of a functional relationship $u = f(x,y)$ is a surface in three-dimensional space, where (x,y) varies over all the points of its domain of definition and where u is the height of the surface above, or below, the point (x,y) in E^2.

Of some special interest are functions which are called analytic.

Definition 1.31. $u = f(x,y)$ is said to be analytic at (x_0,y_0) if and only if $f(x,y)$ can be represented by an infinite series

$$\sum_{i+j=0}^{\infty} a_{i,j}(x - x_0)^i(y - y_0)^j, \qquad i, j \text{ nonnegative integers,}$$

which converges at every point of some circular neighborhood of (x_0,y_0).

Example. $u = e^{x+y}$ is analytic at $(0,0)$ since

$$e^{x+y} \equiv \sum_{i+j=0}^{\infty} \frac{x^i y^j}{i!\,j!}, \qquad i, j \text{ nonnegative integers,}$$

and the indicated series converges at each point of every circular neighborhood of the origin.

Definition 1.32. Let $u = f(x,y)$ be defined on a plane point set M. Then $\lim\limits_{(x,y)\to(x_0,y_0)} f(x,y) = L$, where (x_0,y_0) is a limit point of M and L is a finite constant, means that given $\epsilon > 0$, there exists positive δ such that for all $(x,y) \in M$ which satisfy

$$0 < (x - x_0)^2 + (y - y_0)^2 < \delta,$$

it follows that $|f(x,y) - L| < \epsilon$.

Note that Definition 1.32 includes the concept of a one-sided limit.

Definition 1.33. Let $u = f(x,y)$ be defined on point set M. Then if $(x_0,y_0) \in M$, u is said to be continuous at (x_0,y_0) if and only if, given $\epsilon > 0$, there exists positive δ such that for all $(x,y) \in M$ which satisfy $(x - x_0)^2 + (y - y_0)^2 < \delta$, it follows that

$$|f(x,y) - f(x_0,y_0)| < \epsilon.$$

It is well known that if a function (of any number of variables) is continuous on a closed and bounded point set M, then it is bounded on M, assumes its greatest lower bound at some point of M, and assumes its least upper bound at some point of M. These properties of continuous functions will now be applied to the introduction of various distance functions associated with plane point sets.

Definition 1.34. Let M be a point set in E^2. Let point P_1, with coordinates (x_1,y_1), and point P_2, with coordinates (x_2,y_2), be arbitrary elements of M. Then, provided it exists, the diameter D of set M is defined to be

$$D(M) = \lub_{P_1 \in M, P_2 \in M} \sqrt{(x_2 - x_1)^2 + (y_2 - y_1)^2}. \tag{1.2}$$

Example 1. Let M be the set of points (x,y) whose coordinates satisfy $x^2 + y^2 \le 1$. Then $D(M) = 2$.

Example 2. Let M be the set of points (x,y) whose coordinates satisfy $0 < x < 1, 0 < y < 2$. Then $D(M) = \sqrt{5}$.

Note that if, in E^2, M is nonempty, closed, and bounded, then $\sqrt{(x_2 - x_1)^2 + (y_2 - y_1)^2}$ is continuous on a closed and bounded set. Thus, by a property of continuous functions, it follows that there exist points $\bar{P}_1$, with coordinates $(\bar{x}_1,\bar{y}_1)$, and $\bar{P}_2$, with coordinates $(\bar{x}_2,\bar{y}_2)$, which are both elements of M, such that

$$D(M) = \sqrt{(\bar{x}_2 - \bar{x}_1)^2 + (\bar{y}_2 - \bar{y}_1)^2}.$$

Definition 1.35. Let M_1, M_2 be two sets in E^2. Let P_1, with coordinates (x_1,y_1), be an arbitrary element of M_1, and let P_2, with coordinates (x_2,y_2), be an arbitrary element of M_2. Then, provided it exists, the distance d between the two sets M_1, M_2 is defined to be

$$d(M_1,M_2) = \underset{P_1 \in M_1, P_2 \in M_2}{\text{glb}} \sqrt{(x_2 - x_1)^2 + (y_2 - y_1)^2}. \qquad (1.3)$$

Example 1. Let M_1 be the set of all points (x,y) whose coordinates satisfy $x^2 + y^2 \leq 1$ and let M_2 be the set of all points (x,y) whose coordinates satisfy $x \geq 2$. Then $d(M_1,M_2) = 1$.

Example 2. Let M_1 consist of the single point $(0,0)$ and let M_2 be the set of all points (x,y) whose coordinates satisfy $y = x^2 + 2$. Then $d(M_1,M_2) = 2$.

Note again that if M_1, M_2 are nonempty, closed, and bounded, then by a property of continuous functions, there exists a point $\bar{P}_1$, with coordinates $(\bar{x}_1,\bar{y}_1)$, in M_1, and a point $\bar{P}_2$, with coordinates $(\bar{x}_2,\bar{y}_2)$, in M_2, such that

$$d(M_1,M_2) = \sqrt{(\bar{x}_2 - \bar{x}_1)^2 + (\bar{y}_2 - \bar{y}_1)^2}. \qquad (1.4)$$

Definition 1.36. Let $u = f(x,y)$ be defined on plane point set M. Then, provided the indicated limits exist, the partial derivatives $\dfrac{\partial u}{\partial x}, \dfrac{\partial u}{\partial y}$ are defined by

$$\frac{\partial u}{\partial x} \equiv \lim_{\Delta x \to 0} \frac{f(x + \Delta x, y) - f(x,y)}{\Delta x},$$

$$\frac{\partial u}{\partial y} \equiv \lim_{\Delta y \to 0} \frac{f(x, y + \Delta y) - f(x,y)}{\Delta y}.$$

Note that, by Definition 1.32, one has that the concept of a one-sided derivative is included in Definition 1.36.

Higher-order derivatives are defined in the usual fashion and, in connection with derivatives, the following simple notational devices will often be employed:

$$\left.\begin{array}{l} u_x \equiv \dfrac{\partial u}{\partial x}, \quad u_y \equiv \dfrac{\partial u}{\partial y}, \quad u_{xx} \equiv \dfrac{\partial^2 u}{\partial x^2}, \quad u_{xy} \equiv \dfrac{\partial^2 u}{\partial x\, \partial y}, \quad u_{yy} \equiv \dfrac{\partial^2 u}{\partial y^2}, \\[3mm] u_{xxx} \equiv \dfrac{\partial^3 u}{\partial x^3}, \quad u_{xxy} \equiv \dfrac{\partial^3 u}{\partial x^2\, \partial y}, \ldots . \end{array}\right\} \qquad (1.5)$$

As in the study of functions of one variable, the use of the symbol C^n, where n is a nonnegative integer, will also be convenient. The symbolism "$u \in C^n$" is defined to mean "u and all possible partial derivatives of u up to and including order n are continuous."

Definition 1.37. In E^2, let R be a region, let B be its boundary, and let $f(x,y)$ be continuous on $R \cup B$. Then one defines, in the usual fashion, the double integral by

$$\int\int_R f(x,y)\, d\bar{R} = \lim_{\substack{n \to \infty \\ \max_i[D(\Delta \bar{R}_i)] \to 0}} \sum_{i=1}^{n} f(\alpha_i, \beta_i)\, \Delta S_i,$$

where $\bar{R} = R \cup B$, $\Delta \bar{R}_i$ are the various closed subdivisions of $\bar{R}$, ΔS_i are the areas of the corresponding $\Delta \bar{R}_i$, $(\underline{\alpha}_i, \beta_i) \in \Delta \bar{R}_i$, and $D(\Delta \bar{R}_i)$ are the diameters of the corresponding $\Delta \bar{R}_i$.

In the sequel, double integrals will be considered only for continuous functions while single integrals will be considered for a somewhat more general class of functions.

1.4   Ordinary Differential Equations

Definition 1.38. An ordinary differential equation is an equation which is, or can be put, in the form

$$F\left(x, y, \frac{dy}{dx}, \frac{d^2y}{dx^2}, \ldots, \frac{d^n y}{dx^n}\right) = 0, \tag{1.6}$$

where n is a positive integer and F is a function of the indicated $(n + 2)$ quantities.

Example 1. $\left(\dfrac{dy}{dx}\right)^2 + y^2 = 0$ is an ordinary differential equation.

Example 2. $x^3 \dfrac{d^3y}{dx^3} - x^2 \dfrac{d^2y}{dx^2} + 3x \dfrac{dy}{dx} + y = e^x \sin x$ is an ordinary differential equation.

Definition 1.39. Let M be a set of real numbers. The function $f(x)$, defined on the set M, is said to be a solution of (1.6) on M if and only if $f(x)$ has derivatives up to and including order n on M and

$$F\left(x, f(x), \frac{df(x)}{dx}, \frac{d^2f(x)}{dx^2}, \ldots, \frac{d^n f(x)}{dx^n}\right) \equiv 0$$

on M.

Example 1. If M is the set of all real numbers, then $f(x) = \sin x$ is a solution of

$$\frac{d^2y}{dx^2} + y = 0$$

on M, since $dy/dx = \cos x$, $d^2y/dx^2 = -\sin x$, and

$$-\sin x + \sin x \equiv 0$$

for all real x.

Example 2.   Let M be the set of real numbers x which satisfy $0 < x < 1$.
Then $y = \dfrac{x^{3/2}}{\sqrt{1-x}}$ is a solution on M of

$$2x^3 \frac{dy}{dx} - y(y^2 + 3x^2) = 0$$

since

$$\frac{dy}{dx} = \frac{3x^{1/2}}{2\sqrt{1-x}} + \frac{x^{3/2}}{2(1-x)^{3/2}}$$

and

$$2x^3\left[\frac{3x^{1/2}}{2\sqrt{1-x}} + \frac{x^{3/2}}{2(1-x)^{3/2}}\right] - \frac{x^{3/2}}{\sqrt{1-x}}\left[\frac{x^3}{1-x} + 3x^2\right] \equiv 0$$

on M.

With a view toward studying partial differential equations, we now shall concern ourselves with the study of the solution on the set of all real numbers of the ordinary differential equations

$$\frac{dy}{dx} + ky = 0, \tag{1.7}$$

and

$$\frac{d^2y}{dx^2} + ky = 0, \tag{1.8}$$

where k is an arbitrary constant.

Assuming no results from the theory of ordinary differential equations, we resort to the rudimentary, often fecund, device of trial and error.   Solutions might be sought first from the class of elementary functions.   In particular, consider a trial-and-error solution of the form

$$f(x) = ce^{mx}, \tag{1.9}$$

where c, m are constants to be determined.

Substitution of (1.9) into (1.7) yields

$$mce^{mx} + cke^{mx} \equiv 0,$$

or, equivalently, that

$$c(m + k) = 0.$$

Setting $m = -k$ and allowing c to be arbitrary, however, satisfies this latter identity and reduces (1.9) to

$$f(x) = ce^{-kx}, \tag{1.10}$$

which, it readily follows, satisfies (1.7).   For reasons which are of interest in the theory of ordinary differential equations (see Ref. 14, pp. 43–44), but which are of little significance in the study of partial differential equations, and hence need not concern us, the solution (1.10) is called the *general* solution of (1.7).

Definition 1.40.   If k is an arbitrary, but fixed, real number, then the general solution, for all real x, of

$$\frac{dy}{dx} + ky = 0$$

is defined to be

$$f(x) = ce^{-kx},$$

where c is an arbitrary constant.

Example.   The general solution of $y' - 4y = 0$ is $f(x) = ce^{4x}$, where c is an arbitrary constant.

With respect to (1.8), note that if $k = 0$ the equation can be resolved almost immediately by quadratures to yield $y = c_1 + c_2 x$, where c_1, c_2 are arbitrary constants.   This case is, however, of minor consequence in the study of partial differential equations.   Of deeper significance are the two cases $k < 0, k > 0$.   Suppose first, then, with respect to (1.8), that $k < 0$.   Substitution of (1.9) and its second derivative into (1.8) yields the identity

$$ce^{mx}(m^2 + k) \equiv 0. \tag{1.11}$$

Setting $m^2 + k = 0$ yields, since $k < 0$, the two real roots $m_1 = \sqrt{-k}$, $m_2 = -\sqrt{-k}$.   For these values of m, (1.11) will be satisfied identically without imposing any conditions on the constant c. These results suggest consideration as solutions of (1.8) the two functions

$$f_1(x) = c_1 e^{\sqrt{-k}x} \tag{1.12}$$

$$f_2(x) = c_2 e^{-\sqrt{-k}x}, \tag{1.13}$$

where c_1, c_2 are arbitrary constants.   Simple calculations reveal that (1.12), (1.13) are solutions of (1.8).   However, since c_1, c_2 are arbitrary, it follows that (1.12) and (1.13) are special cases of

$$f(x) = c_1 e^{\sqrt{-k}x} + c_2 e^{-\sqrt{-k}x}, \tag{1.14}$$

where, to obtain (1.12) let $c_2 = 0$, and, to obtain (1.13) let $c_1 = 0$. Further simple calculation reveals that (1.14) is also a solution of (1.8).   Again, for reasons which are of interest in the theory of ordinary differential equations (see Ref. 14, pp. 43–44), but which are of little significance in the study of partial differential equations, and hence need not concern us, the solution (1.14) is called the *general* solution of (1.8).

Definition 1.41.   The general solution, for all real x, of

$$\frac{d^2y}{dx^2} + ky = 0, \qquad k < 0, \tag{1.15}$$

is defined to be

$$f(x) = c_1 e^{\sqrt{-k}x} + c_2 e^{-\sqrt{-k}x} \tag{1.16}$$

where c_1, c_2 are arbitrary constants.

Example.   The general solution of $d^2y/dx^2 - 4y = 0$ is $f(x) = c_1 e^{2x} + c_2 e^{-2x}$, where c_1, c_2 are arbitrary constants.

Consider next, with respect to (1.8), $k > 0$.   With the aid of the complex-valued exponential function, the case $k > 0$ can be treated by considering the results for the case $k < 0$.   Assuming no familiarity with complex functions, we must, however, continue in the simple-minded fashion used for the case $k < 0$.

Trial and error with elementary functions leads to a consideration of a solution of the type

$$f(x) = c \sin mx, \tag{1.17}$$

where c, m are constants to be determined.   Substitution of (1.17) and its second derivative into (1.8) yields

$$-c(\sin mx)(m^2 - k) \equiv 0. \tag{1.18}$$

Setting $m^2 - k = 0$ yields, since $k > 0$, the two real roots $m_1 = \sqrt{k}$, $m_2 = -\sqrt{k}$.   For these values of m, (1.18) will be satisfied identically without imposing any conditions on constant c.   These results suggest consideration of

$$f_1(x) = c_1 \sin \sqrt{k}x, \tag{1.19}$$

where c_1 is an arbitrary constant, as a solution of (1.8).   Direct calculation reveals that (1.19) is a solution of (1.8).   Note that no advantage is realized by also considering $f_2(x) = c_2 \sin (-\sqrt{k}x)$, where c_2 is an arbitrary constant, as a second solution of (1.8) since $f_2(x) = c_2 \sin (-\sqrt{k}x) \equiv -c_2 \sin \sqrt{k}x$ is essentially the same as (1.19).

In an analogous fashion, however,

$$f_2(x) = c_2 \cos \sqrt{k}x, \tag{1.20}$$

where c_2 is an arbitrary constant, can be shown to be a second solution of (1.8) which is essentially different from (1.19). Finally note that (1.19) and (1.20) are special cases of

$$f(x) = c_1 \sin \sqrt{k}x + c_2 \cos \sqrt{k}x, \qquad (1.21)$$

which, for c_1, c_2 arbitrary, is also a solution of (1.8).

Definition 1.42. The general solution, for all real x, of

$$\frac{d^2y}{dx^2} + ky = 0, \qquad k > 0, \qquad (1.22)$$

is defined to be

$$f(x) = c_1 \sin \sqrt{k}x + c_2 \cos \sqrt{k}x, \qquad (1.23)$$

where c_1, c_2 are arbitrary constants.

Example. The general solution for all real x of $\dfrac{d^2y}{dx^2} + 4y = 0$ is $f(x) = c_1 \sin 2x + c_2 \cos 2x$, where c_1, c_2 are arbitrary constants.

1.5 Remarks on Complex-valued Functions

As was noted in the previous section, the fundamental, but tedious, process of trial and error, can often be minimized by a knowledge of complex-valued functions. A facility with such functions often allows such simplification in presentation that some useful ideas and results concerning complex numbers and functions will now be presented. It is to be emphasized, however, that the material of this section will be utilized in this book only as a means of simplifying

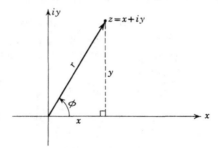

DIAGRAM 1.5

the study of certain real-valued functions associated with partial differential equations.

The quantity $z = x + iy$ (x,y real, $i^2 = -1$) may be considered *algebraically* as a number with the usual rules associated with complex numbers, *geometrically* as a point in the complex z plane, or *physically* as a vector emanating from the origin of the complex z plane and terminating at the point z (consult Diagram 1.5).

If one introduces polar coordinates by the transformation $x = r \cos \phi$, $y = r \sin \phi$ ($-\pi < \phi \leq \pi$, $r > 0$), then any point $z = x + iy$, different from the origin, may be written uniquely in the polar form

$$z = r(\cos \phi + i \sin \phi). \tag{1.24}$$

Example. The polar form for $z = 1 + i$ is

$$z = \sqrt{2}\left(\cos \frac{\pi}{4} + i \sin \frac{\pi}{4}\right).$$

Theorem 1.7.   Theorem of DeMoivre. For n a positive integer

$$(\cos \phi + i \sin \phi)^n \equiv \cos n\phi + i \sin n\phi. \tag{1.25}$$

Proof. The theorem is established by mathematical induction. Note first that (1.25) is valid for $n = 1$. Assume then that (1.25) is valid for $n = k$, that is

$$(\cos \phi + i \sin \phi)^k \equiv \cos k\phi + i \sin k\phi. \tag{1.26}$$

One need only show, then, to complete the proof that (1.25) is valid for $n = k + 1$, that is,

$$(\cos \phi + i \sin \phi)^{k+1} \equiv \cos (k + 1)\phi + i \sin (k + 1)\phi.$$

By use of (1.26), then, one has

$$
\begin{aligned}
(\cos \phi + i \sin \phi)^{k+1} &\equiv (\cos \phi + i \sin \phi)^k(\cos \phi + i \sin \phi) \\
&\equiv (\cos k\phi + i \sin k\phi)(\cos \phi + i \sin \phi) \\
&\equiv (\cos k\phi \cos \phi - \sin k\phi \sin \phi) \\
&\quad + i(\cos k\phi \sin \phi + \sin k\phi \sin \phi) \\
&\equiv \cos (k\phi + \phi) + i \sin (k\phi + \phi) \\
&\equiv \cos (k + 1)\phi + i \sin (k + 1)\phi,
\end{aligned}
$$

and the theorem is proved.

Example. $[\cos (\pi/4) + i \sin (\pi/4)]^8 = \cos 2\pi + i \sin 2\pi = 1$.

Definition 1.43. The absolute value of a complex number $z = x + iy$, denoted $|z|$, is defined by $|z| = \sqrt{x^2 + y^2}$.

Examples. $|1 + i| = \sqrt{2}$, $|3 - 4i| = 5$, $|-7| = 7$, $|0| = 0$, $|2i| = 2$.

Note that if $|z| = 0$, then $x = y = 0$. Also note that if one views complex number z as a vector, then $|z|$ denotes the magnitude of that vector.

Definition 1.44. If to every element z of a nonempty set of complex numbers M_1, there corresponds by some rule a unique element w of a set of complex numbers M_2, then a complex-valued function f is said to be defined. The element w which corresponds under the given rule to the element z is called the value of f at z and one writes $w = f(z)$.

Example 1. $w = 1 + 2z + z^3$ is a complex function defined for all z.
Example 2. $w = 2i + |z|^2$ is a complex function defined for all z.

Note that whenever one wishes to designate that a function under consideration is complex, the modifiers *complex*, or *complex-valued*, will be included in the discussion. The unmodified term *function*, as assumed previously, implies *real-valued function*.

A complex function with extensive application is the Euler function e^{iy}. Before defining the Euler function precisely, let us explore the consequences of assuming that e^{iy}, for all real y, possesses certain desirable properties. Since $e^y \equiv \sum_{n=0}^{\infty} \frac{y^n}{n!}$, for all real y, one might ask that the analogous series $e^{iy} \equiv \sum_{n=0}^{\infty} \frac{(iy)^n}{n!}$ be a valid representation of the given function for all real y. Assuming the convergence of all series under consideration and allowing free reorganization of terms, one would have that for all real y

$$e^{iy} \equiv \sum_{n=0}^{\infty} \frac{(iy)^n}{n!} \equiv \sum_{n=0}^{\infty} \frac{(-1)^n y^{2n}}{(2n)!} + i \sum_{n=0}^{\infty} \frac{(-1)^n y^{2n+1}}{(2n+1)!} \equiv \cos y + i \sin y.$$

Indeed, the properties described above will be most desirable so that one is led to the following definition.

Definition 1.45. The complex-valued Euler function e^{iy} is defined for all real y by

$$e^{iy} \equiv \cos y + i \sin y.$$

Example. $e^{i\pi} = \cos \pi + i \sin \pi = -1$.

Definition 1.46. Let M be a set of complex numbers and let $\{g_n(z)\}$ be a sequence of complex-valued functions defined on M. Let $g(z)$ be a complex-valued function defined on M. Then

$$\lim_{n \to \infty} g_n(z) = g(z)$$

means that for given $z \in M$ and $\epsilon > 0$, there exists an integer $N(\epsilon,z)$ such that for all $n > N$ one has $|g_n(z) - g(z)| < \epsilon$.

Example. Let M be the set of complex numbers z which have the property that $|z| < 1$. Then for all $z \in M$, let $g_n(z) = z^n$. Then

$$\lim_{n \to \infty} g_n(z) = \lim_{n \to \infty} z^n = 0, \qquad |z| < 1. \tag{1.27}$$

For, if $z = 0$, then (1.27) is valid. If $z \neq 0$, consider z in polar form $z = r(\cos \phi + i \sin \phi)$. Since $0 < |z| < 1$, it follows that $0 < r < 1$. Hence

$$\lim_{n \to \infty} g_n(z) = \lim_{n \to \infty} z^n$$
$$= \lim_{n \to \infty} [r(\cos \phi + i \sin \phi)]^n$$
$$= \lim_{n \to \infty} [r^n(\cos \phi + i \sin \phi)^n]$$
$$= \lim_{n \to \infty} [r^n(\cos n\phi + i \sin n\phi)].$$

However,

$$|r^n(\cos n\phi + i \sin n\phi)| = r^n,$$

so that

$$\lim_{n \to \infty} |r^n(\cos n\phi + i \sin n\phi)| = \lim_{n \to \infty} r^n = 0,$$

which, from Definition 1.46 and the above discussion, implies that

$$\lim_{n \to \infty} g_n(z) = \lim_{n \to \infty} z^n = \lim_{n \to \infty} [r^n(\cos n\phi + i \sin n\phi)] = 0.$$

Definition 1.47. Let M be a set of complex numbers and let $\{f_n(z)\}$ be a sequence of complex-valued functions defined on M. Then the kth partial sum $S_k(z)$ of the series $\sum_{n=1}^{\infty} f_n(z)$ is defined for all $z \in M$ by

$$S_k(z) = \sum_{n=1}^{k} f_n(z). \tag{1.28}$$

Example. Let M be the set of all complex numbers z which possess the property $|z| < 1$. Let $f_1(z) = 1, f_n(z) = z^{n-1}$; $n = 2, 3, \ldots$. Then for all $z \in M$, relative to $\sum_{n=1}^{\infty} f_n(z)$, one has

$$S_k(z) = \frac{1 - z^k}{1 - z}; \qquad k = 1, 2, 3, \ldots. \tag{1.29}$$

Definition 1.48. Let M be a set of complex numbers and let $\{f_n(z)\}$ be a sequence of complex-valued functions each of which is defined on M. Let $f(z)$ be a complex-valued function defined on

M.   Then the series $\sum_{n=1}^{\infty} f_n(z)$ is said to converge to $f(z)$ on M, or to have sum $f(z)$ on M, if and only if for each $z \in M$

$$\lim_{k \to \infty} S_k(z) = f(z),$$

where $S_k(z)$ is given by (1.28). If $\sum_{n=1}^{\infty} f_n(z)$ has sum $f(z)$ on set M, one writes $\sum_{n=1}^{\infty} f_n(z) \equiv f(z)$, $z \in M$.

Example.

$$1 + z + z^2 + z^3 + \cdots + z^{n-1} + \cdots \equiv \frac{1}{1-z}\,; \qquad |z| < 1. \quad (1.30)$$

For let M be the set of complex numbers z which possess the property $|z| < 1$.   Let $f_1(z) = 1$, $f_n(z) = z^{n-1}$; $n = 2, 3, 4, \ldots$.   Then, for all $z \in M$, the kth partial sum of the series $\sum_{n=1}^{\infty} f_n(z)$ is given by (1.29).   Since $|z| < 1$, it follows from (1.27) and Definition 1.26 that

$$\lim_{k \to \infty} S_k(z) = \frac{1}{1-z}\,,$$

and (1.30) is established.

EXERCISES

1.1. Give ten examples of sets.

1.2. Find $A \cup B$ and $A \cap B$ for each of the following:

(a) A is the set of integers, B is the set of nonnegative integers. (*Ans.* $A \cup B = A$, $A \cap B = B$.)

(b) A is the set of positive rational numbers, B is the set of nonnegative rational numbers.

(c) A is the set of real numbers, B is the set of rational numbers. (*Ans.* $A \cup B = A$, $A \cap B = B$.)

(d) A is the set of algebraic irrationals, B is the set of transcendental irrationals. (*Ans.* $A \cup B$ is the set of irrationals, $A \cap B = \theta$.)

1.3. Given two sets A, B, prove that $(A \cap B) \subset A$ and $(A \cap B) \subset B$.

1.4. Given two sets A, B, prove that $A \subset (A \cup B)$ and $B \subset (A \cup B)$.

1.5. Given three sets A, B, D, prove that if $A \subset B$ and $B \subset D$, then $A \subset D$.

1.6. Two sets A and B are said to be in one-to-one correspondence if and only if for every element of A there corresponds, by some rule, a unique element of B and for every element of B there corresponds, by some rule, a unique element of A.   Prove that

(a) if A is the set of positive integers and B is the set of nonnegative integers, then A and B can be put in one-to-one correspondence.

(b) if A is the set of positive even integers and B is the set of positive odd integers, then A and B can be put in one-to-one correspondence.

(c) if A is the set of positive integers and B is the set of integers, then A and B can be put in one-to-one correspondence.

1.7. In E^2, describe geometrically each set M, where M is the set of all points (x,y) whose coordinates satisfy:

(a) $x^2 + y^2 < 1$
(b) $x^2 + y^2 \leq 1$
(c) $x^2 + y^2 > 0$
(d) $x^2 + y^2 \geq 1$
(e) $1 < x^2 + y^2 < 2$
(f) $1 < x^2 + y^2 \leq 2$
(g) $1 \leq x^2 + y^2 \leq 2$
(h) $0 < x < 2$, y arbitrary
(i) $-1 \leq y \leq 3$, x arbitrary

(j) $0 \leq x \leq 1$, $\quad y \geq 0$
(k) $0 < y \leq 2$, $\quad x < 0$
(l) $y < x + 1$
(m) $y \geq 2x - 3$
(n) $x > 0$, $\quad y > 0$
(o) $x \leq 0$, $\quad y > 0$
(p) $x^2 + 2y^2 \leq 4$
(q) $x^2 < y$
(r) $x^2 < y^2 + 1$.

1.8. For each set of Exercise 1.7, determine $E^2 - M$.

1.9. For each set of Exercise 1.7, determine whether or not M is (a) open, (b) closed, (c) bounded, (d) connected, (e) simply connected, (f) a region, (g) a closed region.

1.10. Determine the boundary of each set M of Exercise 1.7.

1.11. In E^2, let M_1 be the set of all points (x,y) whose coordinates satisfy $(x - 2)^2 + y^2 < 1$ and let M_2 be the set of all points (x,y) whose coordinates satisfy $(x + 2)^2 + y^2 < 1$. Let $M = M_1 \cup M_2$. Determine whether or not M is (a) open, (b) bounded, (c) connected, (d) simply connected, (e) a region, (f) a closed region. (*Ans.* M is open and bounded. M is not connected, simply connected, a region, or a closed region.)

1.12. For each of the following, give an example, if possible, of a plane point set M and a point P such that

(a) P is a limit point of M, but P is not a boundary point of M,
(b) P is a boundary point of M, but P is not a limit point of M,
(c) P is a limit point of M, P is not an element of M, and M is closed.

1.13. For each of the following, give an example of a plane point set M which is:

(a) open and closed
(b) neither open nor closed
(c) connected, but neither open nor closed
(d) closed and not bounded
(e) closed and bounded, but not connected.

1.14. Show that in E^2 any set M which consists of a finite number of points is a closed set.

1.15. In E^2, a closed, bounded set M is said to be convex if and only if, given any two points P_1, P_2 of M, then the entire straight-line segment which has P_1, P_2 for end points is also contained in M. Show that in E^2, if a closed, bounded set M is convex, and if point P is not an element of M, then there is at least one straight line L which passes through P but has no points in common with M.

1.16. For each of the following parametric representations, sketch in E^2 the bounded curve which is so defined and determine whether or not each such curve is simple, closed, simple and closed, smooth, and/or a contour:

(a) $x = 1$, $y = t$, $\quad 0 \leq t \leq 2\pi$
(b) $x = t$, $y = 2$, $\quad 0 \leq t \leq 2\pi$
(c) $x = t$, $y = -2t$, $\quad 0 \leq t \leq 2\pi$

(d) $x = t, y = t^2, \quad 0 \le t \le 2\pi$
(e) $x = t^2, y = \sin t, \quad 0 \le t \le 2\pi$
(f) $x = e^t, y = \cos t, \quad 0 \le t \le 2\pi$
(g) $x = \log (1 + t), y = \cos t, \quad 0 \le t \le 2\pi$ (Note: $\log m = \log_e m$.)
(h) $x = \cos^2 t, y = \sin t, \quad 0 \le t \le 2\pi$
(i) $x = t - \sin t, y = 1 - \cos t, \quad 0 \le t \le 2\pi$.

1.17. Give five examples of curves in E^2 which are not bounded curves.

1.18. Find the slope of the tangent line to each curve of Exercise 1.16 at the point (x,y) which corresponds to parameter value $t = \pi$. (*Ans.* (a) undefined; (b) 0; (d) 2π; (e) $-\dfrac{1}{2\pi}$; (i) 0.)

1.19. Each of the following functions is defined on E^2 and defines a surface. Draw each surface.

(a) $u = x - y$ (d) $u = x^2 + y^2$
(b) $u = x + y^2$ (e) $u = x^2 - y^2$.
(c) $u = x - y^3$

1.20. Find the equation of the tangent plane to the surface defined by $u = x^2 - y^2$ at that point of the surface corresponding to $x = 1$, $y = 1$. (*Ans.* $u = 2x - 2y$.)

1.21. Show that if each of the indicated functions is defined at all points of E^2 except (0,0), then:

(a) $\displaystyle \lim_{(x,y) \to (1,1)} \frac{2x^3 - y^3}{x^2 + y^2} = \frac{1}{2}$

(b) $\displaystyle \lim_{(x,y) \to (0,0)} \frac{2x^3 - y^3}{x^2 + y^2} = 0$

(c) $\displaystyle \lim_{(x,y) \to (0,0)} \frac{x^2 - y^2}{x^2 + y^2}$ does not exist

(d) $\displaystyle \lim_{(x,y) \to (0,0)} \frac{xy(x^2 - y^2)}{x^2 + y^2} = 0$.

1.22. Let $M = E^2$ and define $u = f(x,y)$ at all points of M, except at (0,0), by

$$f(x,y) = \frac{\sin (x^2 + y^2)}{x^2 + y^2}.$$

Show how to define $f(x,y)$ at (0,0) so that the function is continuous at (0,0).

1.23. For each of the following functions, defined on E^2, find $f_x, f_y, f_{xx}, f_{xy}, f_{yx}, f_{yy}$ at the point (0,0):

(a) $f(x,y) = \sin xy - e^{x+y}$
(b) $f(x,y) = x^3 y - xy^2 + x \cos y$ (*Ans.* 1,0,0,0,0,0)
(c) $f(x,y) = \sin (e^{x^2+y^2})$
(d) $f(x,y) = e^{xy} + e^{x+y}$ (*Ans.* 1,1,1,2,2,1)
(e) $f(x,y) = \sin^2 xy + \cosh (x^2 - y^2)$.

1.24. Let $f(x,y)$ be defined on a plane point set M. Let $(x,y) \in M$. Give a set of sufficient conditions to ensure that $f_{xy} \equiv f_{yx}$ at (x,y).

1.25. In E^2, let M be the set of points (x,y) which possess the property that $x^2 + y^2 \leq 4$. Then evaluate

(a) $\displaystyle \iint_M (x^2 + y^2)\, dM$ (*Ans.* 8π)

(b) $\displaystyle \iint_M xy\, dM.$

1.26. Show that $u = (1 - x - y)^{-1}$ is analytic at any point (x,y) whose coordinates satisfy $|x| + |y| < 1$.

1.27. For each of the following, find the diameter D of the set M, where M is the set of all points (x,y) of E^2 whose coordinates satisfy

(a) $x^2 + y^2 < 1$ (*Ans.* 2)
(b) $x^2 + y^2 \leq 1$ (*Ans.* 2)
(c) $1 \leq x^2 + y^2 \leq 4$ (*Ans.* 8)
(d) $1 \leq x \leq 2, 2 \leq y \leq 4.$ (*Ans.* $\sqrt{3}$)

1.28. For each of the plane point sets of Exercise 1.27, find, if possible, a pair of points P_1, P_2 which are elements of the given set and such that the distance between the two points coincides with the diameter of the set.

1.29. Find the distance d between the pairs of sets M_1, M_2 of E^2 where

(a) M_1 is the set of all points (x,y) whose coordinates satisfy the property $x^2 + y^2 < 1$, M_2 is the single point $(1,1)$. (*Ans.* $\sqrt{2} - 1$.)

(b) M_1 is the set of all points (x,y) whose coordinates satisfy the property $x^2 + y^2 \leq 1$, M_2 is the single point $(1,1)$. (*Ans.* $\sqrt{2} - 1$.)

(c) M_1 is the set of all points (x,y) whose coordinates satisfy the property $x^2 + y^2 \leq 1$, M_2 is the set of all points (x,y) whose coordinates satisfy the properties $-1 \leq x \leq 1, 2 \leq y \leq 4$.

(d) M_1 is the set of all points (x,y) whose coordinates satisfy the property $x^2 + y^2 \leq 1$, M_2 is the set of all points (x,y) whose coordinates satisfy the properties $3 \leq x \leq 4, 3 \leq y \leq 4$.

1.30. For each pair of plane point sets M_1, M_2 given in Exercise 1.29, find, if possible, points $P_1 \in M_1$, $P_2 \in M_2$ such that the distance between P_1 and P_2 coincides with the distance between M_1 and M_2.

1.31. Verify whether or not each of the given functions is a solution of the accompanying ordinary differential equation for the given values of x:

(a) $y = x - \pi$, $\dfrac{d^3y}{dx^3} - 3\dfrac{d^2y}{dx^2} + \dfrac{dy}{dx} - 1 = 0$, $-1 < x < 1$

(b) $y = \frac{1}{2}(x^3 + x^2 - 3x - 1) + \frac{1}{6}e^{-2x} - \frac{1}{7}\cos 3x$, $y'' + 2y = x^3 + x^2 + e^{-2x} + \cos 3x$, $-1 \leq x \leq 1$

(c) $y = e^x + e^{-x} + \sin x + \cos x$, $\dfrac{d^8y}{dx^8} - y = 0$, $x > 0$

(d) $y = e^{2x} + \frac{1}{10}\sin x + \frac{3}{10}\cos x$, $y'' - 3y' + 2y = \sin x$, $x \geq 0.$

1.32. Find the general solution for all real x of each of the following ordinary differential equations:

(a) $y' + 4y = 0$ (*Ans.* $y = ce^{-4x}$)

(b) $y' - 4y = 0$ (*Ans.* $y = ce^{4x}$)

(c) $y' + 9y = 0$ (*Ans.* $y = ce^{-9x}$)

(d) $y' + \sqrt{2}y = 0$ (*Ans.* $y = ce^{-\sqrt{2}x}$)

(e) $y' - \pi y = 0$

(f) $y'' + 4y = 0$ (*Ans.* $y = c_1 \sin 2x + c_2 \cos 2x$)

(g) $y'' - 4y = 0$ (*Ans.* $y = c_1 e^{2x} + c_2 e^{-2x}$)

(h) $y'' + 9y = 0$ (*Ans.* $y = c_1 \sin 3x + c_2 \cos 3x$)

(i) $y'' - 9y = 0$ (*Ans.* $y = c_1 e^{3x} + c_2 e^{-3x}$)

(j) $y'' + 3y = 0$ (*Ans.* $y = c_1 \sin \sqrt{3}x + c_2 \cos \sqrt{3}x$)

(k) $y'' - 3y = 0$ (*Ans.* $y = c_1 e^{\sqrt{3}x} + c_2 e^{-\sqrt{3}x}$)

(l) $\frac{1}{2}y'' + 5y = 0$.

1.33. For each complex number $z = x + iy$, find $|z|$ and the polar form of z:

(a) $z = 1 + i$ (*Ans.* $\sqrt{2}$, $\sqrt{2}\,[\cos(\pi/4) + i\sin(\pi/4)]$)

(b) $z = 3 - 4i$

(c) $z = -7$ (*Ans.* $7, 7(\cos \pi + i \sin \pi)$)

(d) $z = 4i$.

1.34. If $z_1 = x_1 + iy_1$, $z_2 = x_2 + iy_2$ are two complex numbers, prove that

(a) $|z_1 + z_2| \le |z_1| + |z_2|$

(b) $|z_1 - z_2| \ge |z_1| - |z_2|$

(c) $|z_1 + z_2| \ge ||z_1| - |z_2||$

(d) $|z_1 - z_2| \ge ||z_1| - |z_2||$

(e) $|z_1 z_2| = |z_1| \cdot |z_2|$

(f) $\dfrac{|z_1|}{|z_2|} = \left|\dfrac{z_1}{z_2}\right|$, provided $z_2 \ne 0$.

1.35. If $z = x + iy$ is a complex number, prove that $|z| \ge \dfrac{|x| + |y|}{\sqrt{2}}$.

1.36. Simplify each of the following by DeMoivre's theorem:

(a) $\left(\cos \dfrac{\pi}{3} + i \sin \dfrac{\pi}{3}\right)^3$ (*Ans.* -1)

(b) $\left(\cos \dfrac{\pi}{4} + i \sin \dfrac{\pi}{4}\right)^8$ (*Ans.* 1)

(c) $\left(\cos \dfrac{\pi}{8} + i \sin \dfrac{\pi}{8}\right)^4$ (*Ans.* i)

(d) $\left(\cos \dfrac{\pi}{60} + i \sin \dfrac{\pi}{60}\right)^{15}$. $\left(Ans.\ \dfrac{\sqrt{2}}{2} + i\dfrac{\sqrt{2}}{2}\right)$

1.37. For each of the following, find the five complex numbers which satisfy:

(a) $z^5 = 1$

(b) $z^5 = i$.

1.38. If $f(z) = z + i$, find $f(0)$, $f(1)$, $f(i)$, $f(1 + i)$, $f(1 - i)$, $f(3 - 4i)$.
(*Ans.* $i, i + 1, 2i, 1 + 2i, 1, 3 - 3i$.)

1.39. If $f(z) = z^2 - 2z + 1$, find $f(0)$, $f(1)$, $f(-2i)$, $f(1 - i)$, $f(3 - 4i)$.

1.40. Simplify each of the following complex functions:

(*a*) $e^{\pi i/2}$ (*Ans.* i)
(*b*) $e^{\pi i}$ (*Ans.* -1)
(*c*) $e^{2\pi i}$ (*Ans.* 1)
(*d*) $e^{5\pi i/4}$.

1.41. Find the sum of each of the following series:

(*a*) $\sum_{n=1}^{\infty} (\tfrac{2}{3})^{n-1}$ (*Ans.* 3)

(*b*) $\sum_{n=1}^{\infty} (-\tfrac{1}{2})^{n-1}$ (*Ans.* $\tfrac{2}{3}$)

(*c*) $1 + \sum_{n=1}^{\infty} (\tfrac{1}{2} + \tfrac{1}{2}i)^n$ (*Ans.* $1 + i$)

(*d*) $1 + \sum_{n=1}^{\infty} (\tfrac{1}{5} - \tfrac{1}{5}i)^n$.   (*Ans.* $\tfrac{50}{29} - \tfrac{20}{29}i$)

1.42. Show that for all complex numbers z for which $|z| > 1$, the series $1 + \sum_{n=1}^{\infty} z^n$ does *not* converge.

Chapter 2

FOURIER SERIES

One of the most powerful concepts available for constructing solutions of differential equations is that of an infinite series, and, in light of the ubiquitous problems which display aspects of a periodic and/or a discontinuous nature, those infinite series known as Fourier series attain a place of special importance.

2.1 Trigonometric and Fourier Series

Definition 2.1. A real-valued function of one variable, $F(x)$, defined for all real x, is said to be periodic and to have period p if and only if for fixed $p > 0$ and for all x

$$F(x + p) = F(x).$$

Example 1. $\sin kx$, $\cos kx$; $k = 0, 1, 2, \ldots$ are periodic functions.

Example 2. $\sin x + \sin 2x + \sin 3x$ is a periodic function and has period 2π.

Example 3. $\frac{1}{2} + \cos x + \cos 2x + \cdots + \cos nx$ is a periodic function and has period 2π.

Note that if $F(x)$ is periodic with period p, then it is also periodic with period kp; $k = 2, 3, \ldots, n - 1, n$, and, moreover, it is possible, but not necessary, that $F(x)$ is periodic with a period p^*, where $0 < p^* < p$. Also note that the periodic function given in Example 3, above, can be written in a more compact form as described in the following lemma.

Lemma 2.1. For all values of ϕ for which $\sin(\phi/2) \neq 0$,

$$\frac{1}{2} + \sum_{k=1}^{n} \cos k\phi \equiv \frac{\sin(n + \frac{1}{2})\phi}{2\sin(\phi/2)}. \tag{2.1}$$

Proof. Since

$$2 \cos A \sin B \equiv \sin (A + B) - \sin (A - B),$$

letting $A = k\phi$, $B = \phi/2$ yields

$$2 \cos k\phi \sin \frac{\phi}{2} \equiv \sin (k + \tfrac{1}{2})\phi - \sin (k - \tfrac{1}{2})\phi.$$

For $k = 1, 2, \ldots, n$, respectively, this latter identity implies

$$2 \cos \phi \sin \frac{\phi}{2} \equiv \sin \frac{3\phi}{2} - \sin \frac{\phi}{2}$$

$$2 \cos 2\phi \sin \frac{\phi}{2} \equiv \sin \frac{5\phi}{2} - \sin \frac{3\phi}{2}$$

$$2 \cos 3\phi \sin \frac{\phi}{2} \equiv \sin \frac{7\phi}{2} - \sin \frac{5\phi}{2}$$

$$\vdots \qquad\qquad \vdots \qquad\qquad \vdots$$

$$2 \cos n\phi \sin \frac{\phi}{2} \equiv \sin \frac{2n + 1}{2} \phi - \sin \frac{2n - 1}{2} \phi.$$

Summing both sides of the latter set of identities yields

$$2 \sin \frac{\phi}{2} \sum_{k=1}^{n} \cos k\phi \equiv \sin \frac{2n + 1}{2} \phi - \sin \frac{\phi}{2},$$

from which (2.1) readily follows.

Lemma 2.2. If one defines

$$\frac{\sin (n + \tfrac{1}{2})\phi}{2 \sin (\phi/2)} \equiv n + \tfrac{1}{2}, \qquad \phi = 0, \tag{2.2}$$

then (2.1) is a valid identity for all ϕ in the range $-\pi \leq \phi \leq \pi$.

Proof. For $\phi \neq 0$, the proof follows from Lemma 2.1. For $\phi = 0$, one has

$$\frac{1}{2} + \sum_{k=1}^{n} \cos k\phi = \frac{1}{2} + \sum_{k=1}^{n} 1 = \frac{1}{2} + n,$$

and the lemma is proved.

Note then, by Lemmas 2.1 and 2.2, that the function

$$g(\phi) = \begin{cases} \dfrac{\sin (n + \tfrac{1}{2})\phi}{2 \sin (\phi/2)}, & -\pi \leq \phi < 0, \, 0 < \phi \leq \pi \\ n + \tfrac{1}{2}, & \phi = 0 \end{cases}$$

is continuous on $-\pi \leq \phi \leq \pi$.

Definition 2.2. A trigonometric series is an infinite series of the form

$$\alpha + \sum_{n=1}^{\infty} (a_n \cos nx + b_n \sin nx), \tag{2.3}$$

where α, a_n, b_n; $n = 1, 2, 3, \ldots$, are constants.

Example. $\left[1 + \sum_{n=1}^{\infty} (n \cos nx + n^2 \sin nx)\right]$ is a trigonometric series.

Suppose now that $F(x)$ is a periodic, integrable function. For our first considerations, let $F(x)$ have period 2π, so that a description of $F(x)$ for $-\pi \leq x \leq \pi$ is sufficient to describe $F(x)$ for all x. Assume, then, $F(x)$ can be represented for $-\pi \leq x \leq \pi$ by a trigonometric series, so that

$$F(x) \equiv \alpha + \sum_{n=1}^{\infty} (a_n \cos nx + b_n \sin nx), \qquad -\pi \leq x \leq \pi, \tag{2.4}$$

and let us see what can be determined, in a heuristic fashion, concerning the constants α, a_n, b_n; $n = 1, 2, \ldots$.

If termwise integration is valid, then (2.4) implies

$$\int_{-\pi}^{\pi} F(x)\, dx \equiv \int_{-\pi}^{\pi} \alpha\, dx + \int_{-\pi}^{\pi} \left[\sum_{n=1}^{\infty} (a_n \cos nx + b_n \sin nx)\right] dx$$

$$\equiv 2\pi\alpha + \sum_{n=1}^{\infty} \left(a_n \int_{-\pi}^{\pi} \cos nx\, dx + b_n \int_{-\pi}^{\pi} \sin nx\, dx\right)$$

$$\equiv 2\pi\alpha,$$

so that

$$\alpha = \frac{1}{2\pi} \int_{-\pi}^{\pi} F(x)\, dx. \tag{2.5}$$

Multiplying each term of (2.4) by $\cos x$ and assuming convergence implies

$$F(x) \cos x \equiv \alpha \cos x + \sum_{n=1}^{\infty} (a_n \cos nx \cos x + b_n \sin nx \cos x).$$

Assuming the validity of termwise integration then yields

$$\int_{-\pi}^{\pi} F(x) \cos x\, dx \equiv \alpha \int_{-\pi}^{\pi} \cos x\, dx$$

$$+ \sum_{n=1}^{\infty} \left(a_n \int_{-\pi}^{\pi} \cos nx \cos x\, dx + b_n \int_{-\pi}^{\pi} \sin nx \cos x\, dx\right)$$

$$\equiv \sum_{n=1}^{\infty} \left(a_n \int_{-\pi}^{\pi} \cos nx \cos x\, dx + b_n \int_{-\pi}^{\pi} \sin nx \cos x\, dx\right).$$

However, $\int_{-\pi}^{\pi} \cos nx \cos x \, dx = 0$, if $n \neq 1$; and $\int_{-\pi}^{\pi} \sin nx \cos x \, dx$
$= 0$ for all n, so that

$$\int_{-\pi}^{\pi} F(x) \cos x \, dx \equiv a_1 \int_{-\pi}^{\pi} \cos^2 x \, dx \equiv \pi a_1,$$

or, equivalently,

$$a_1 = \frac{1}{\pi} \int_{-\pi}^{\pi} F(x) \cos x \, dx. \qquad (2.6)$$

Repeating the assumptions and the type of argument which led to (2.6), one has, in a more general fashion, that (2.4) implies

$F(x) \cos mx \equiv \alpha \cos mx$
$$+ \sum_{n=1}^{\infty} (a_n \cos nx \cos mx + b_n \sin nx \cos mx), \quad m = 1, 2, \dots,$$
$$(2.7)$$

$F(x) \sin mx \equiv \alpha \sin mx$
$$+ \sum_{n=1}^{\infty} (a_n \cos nx \sin mx + b_n \sin nx \sin mx), \quad m = 1, 2, \dots.$$
$$(2.8)$$

Recalling that

$$\left. \begin{array}{l} \int_{-\pi}^{\pi} \cos mx \cos nx \, dx = \int_{-\pi}^{\pi} \sin mx \sin nx \, dx = 0, \quad m \neq n, \\[2mm] \int_{-\pi}^{\pi} \cos mx \sin nx \, dx = 0 \\[2mm] \int_{-\pi}^{\pi} \cos^2 mx \, dx = \int_{-\pi}^{\pi} \sin^2 mx \, dx = \pi, \quad m \neq 0, \end{array} \right\} \qquad (2.9)$$

and integrating termwise, one finds that (2.7) to (2.9) imply

$$a_n = \frac{1}{\pi} \int_{-\pi}^{\pi} F(x) \cos nx \, dx, \qquad n = 1, 2, \dots, \qquad (2.10)$$

$$b_n = \frac{1}{\pi} \int_{-\pi}^{\pi} F(x) \sin nx \, dx, \qquad n = 1, 2, \dots. \qquad (2.11)$$

Thus, (2.5), (2.10), and (2.11) determine the constants α, a_n, b_n for trigonometric series (2.4). Note that while a_n is defined for $n = 1, 2, \dots$, substitution of $n = 0$ into (2.10) yields

$$a_0 = \frac{1}{\pi} \int_{-\pi}^{\pi} F(x) \, dx = 2\alpha,$$

so that if one wishes to let $\alpha = a_0/2$, (2.10) and (2.11) can then be used to determine all the coefficients of trigonometric series (2.4).

The above discussion has thus led quite naturally to the following definition.

Definition 2.3.   Let $F(x)$ be defined and integrable for $-\pi \leq x \leq \pi$.   Then the trigonometric series

$$\frac{a_0}{2} + \sum_{n=1}^{\infty} (a_n \cos nx + b_n \sin nx), \qquad (2.12)$$

where

$$a_n = \frac{1}{\pi} \int_{-\pi}^{\pi} F(x) \cos nx \, dx, \qquad n = 0, 1, 2, \ldots, \qquad (2.13)$$

$$b_n = \frac{1}{\pi} \int_{-\pi}^{\pi} F(x) \sin nx \, dx, \qquad n = 1, 2, \ldots, \qquad (2.14)$$

is said to be the *Fourier series* associated with $F(x)$.

Example.   Find the Fourier series associated with function $F(x)$ where $F(x)$ is defined for all real x, has period 2π, and $F(x) = |x|$ on $-\pi \leq x \leq \pi$, or, in shorthand notation, where $F(x)$ is defined for all real x by

$$F(x) = \begin{cases} |x|, & -\pi \leq x \leq \pi \\ F(x + 2\pi). \end{cases}$$

Solution.   Since $F(x)$ is defined and integrable for $-\pi \leq x \leq \pi$, (2.13) and (2.14) imply $a_0 = \pi; a_n = (2/\pi n^2)[(\cos n\pi) - 1], n = 1, 2, \ldots; b_n = 0$, $n = 1, 2, \ldots$, so that the associated Fourier series is

$$\frac{\pi}{2} + \sum_{n=1}^{\infty} \frac{2}{\pi n^2} [(\cos n\pi) - 1] \cos nx$$

$$= \frac{\pi}{2} - \frac{4}{\pi} \left(\cos x + \frac{1}{3^2} \cos 3x + \frac{1}{5^2} \cos 5x \cdots \right).$$

The fundamental problem of establishing general conditions under which the Fourier series associated with a function $F(x)$ actually converges and represents $F(x)$ for all real x will now be considered.   Though more general conditions than those to be established can be found, those to be developed will suffice for many practical purposes.

Definition 2.4.   Let a, b, be two real numbers with the property $a < b$.   Then $F(x)$ is said to be sectionally continuous on $a \leq x \leq b$ if and only if:

(a) $F(x)$ is continuous for all, except possibly a finite number of, values of x in $a \leq x \leq b$, and

(b) $F(x)$ has right-hand limit at $x = a$, left-hand limit at $x = b$, and both right- and left-hand limits for all other values of x in $a < x < b$.

Example 1. The function defined by

$$F(x) = \begin{cases} 1, & 0 < x \leq \pi \\ 0, & x = 0 \\ -1, & -\pi \leq x < 0 \end{cases}$$

is sectionally continuous on $-\pi \leq x \leq \pi$.

Example 2. The function defined by

$$F(x) = \begin{cases} 1, & 0 < x < \pi \\ -1, & -\pi < x < 0 \end{cases}$$

is sectionally continuous on $-\pi \leq x \leq \pi$.

Note that a function can be sectionally continuous on $a \leq x \leq b$ and yet not be defined at every point of $a \leq x \leq b$ (see Example 2, above). Observe also that if $F(x)$ is sectionally continuous on $a \leq x \leq b$, then $F(x)$ is bounded and integrable on $a \leq x \leq b$.

With respect to sectionally continuous functions, the following notation will be useful. $F(x_0^+)$ will be used to denote the right-hand limit of $F(x)$ as x goes to x_0. Similarly, $F(x_0^-)$ will be used to denote the left-hand limit of $F(x)$ as x goes to x_0.

Example. Consider, on $-\pi \leq x \leq \pi$, the function defined by

$$F(x) = \begin{cases} 1, & 0 < x \leq \pi \\ 0, & x = 0 \\ -1, & -\pi \leq x < 0. \end{cases}$$

Then $F(0^+) = 1$, $F(0^-) = -1$, $F(1^+) = 1$, $F(1^-) = 1$, $F(-1^+) = -1$, $F(-1^-) = -1$, $F(\pi^-) = 1$, $F(-\pi^+) = 1$, $F(\pi^+)$ and $F(-\pi^-)$ are not defined, $\dfrac{F(0^+) + F(0^-)}{2} = 0$, $\dfrac{F(1^+) + F(1^-)}{2} = 1$.

The following lemma germane to sectionally continuous functions will be of value in later work.

Lemma 2.3. Riemann's Lemma. Let a, b be two real numbers with the property $a < b$. Let M be the set of all real numbers x which satisfy $a \leq x \leq b$. If $F(x)$ is sectionally continuous on M, then

$$\lim_{n \to \infty} \int_a^b F(x) \sin nx \, dx = 0. \tag{2.15}$$

Proof. The proof will be given by considering three cases.

CASE 1. Suppose $F(x)$ is continuous on M. Then the substitution $x = t + h$, where $h = \pi/n$ and n is sufficiently large so that $b - h > a$, transforms the integral

$$I = \int_a^b F(x) \sin nx \, dx \qquad (2.16)$$

into

$$I = -\int_{a-h}^{b-h} F(t + h) \sin nt \, dt.$$

Since t is only a dummy variable, the latter integral may be rewritten

$$I = -\int_{a-h}^{b-h} F(x + h) \sin nx \, dx. \qquad (2.16')$$

Hence, (2.16) and (2.16$'$) imply

$$2I = -\int_{a-h}^{a} F(x + h) \sin nx \, dx + \int_a^{b-h} [F(x) - F(x + h)] \sin nx \, dx$$

$$+ \int_{b-h}^{b} F(x) \sin nx \, dx. \qquad (2.17)$$

Since $F(x)$ is continuous on M, it is bounded on M. For all $x \in M$, then, there exists positive constant A such that

$$|F(x)| < A. \qquad (2.18)$$

Then (2.17) and (2.18) imply

$$2|I| \leq 2Ah + \int_a^{b-h} |F(x) - F(x + h)| \, dx.$$

Since $F(x)$ is uniformly continuous on $a \leq x \leq b$, n can always be selected sufficiently large so that given $\epsilon > 0$, h will also satisfy

$$Ah = \frac{A\pi}{n} < \frac{\epsilon}{2}$$

and so that for all x and $x + h$ for which $a \leq x \leq b - h$ and $a \leq x + h \leq b - h$, respectively, one will have

$$|F(x) - F(x + h)| < \frac{\epsilon}{b - a} .$$

Then

$$|I| \leq Ah + \frac{1}{2} \int_a^{b-h} |F(x) - F(x + h)| \, dx$$

$$< \frac{\epsilon}{2} + \frac{1}{2} \frac{\epsilon}{b - a} (b - a - h) < \epsilon. \qquad (2.19)$$

Since (2.19) is valid for all $\epsilon > 0$ and for all n which satisfy $A\pi/n < \epsilon/2$, the lemma readily follows.

CASE 2. Suppose $F(x)$ is continuous on $a < x < b$. Since $F(x)$ is sectionally continuous on M, define $F^*(x)$ on M by

$$F^*(x) = \begin{cases} F(a^+), & x = a \\ F(x), & a < x < b \\ F(b^-), & x = b. \end{cases}$$

Then $F^*(x)$ is continuous on M, and since

$$\int_a^b F^*(x) \sin nx \, dx = \int_a^b F(x) \sin nx \, dx,$$

it follows from Case 1 that (2.15) is valid.

CASE 3. Suppose $F(x)$ has a finite number of discontinuities on M. Without loss of generality, assume these occur at $x_1, x_2, x_3, \ldots, x_m$, where $x_1 < x_2 < x_3 < \cdots < x_m$. For notational purposes, set $a = x_0$, $b = x_{m+1}$. Let M_i denote the set of all real numbers x which satisfy $x_{i-1} \leq x \leq x_i$, $i = 1, 2, \ldots, m + 1$. Then application of the method of Case 2 to each set M_i, $i = 1, 2, \ldots, m + 1$, readily implies that (2.15) is valid.

Since by Definition 2.4, only Cases 1, 2, 3 are possible, the lemma is proved.

Definition 2.5. $F(x)$ is said to be right-quasi-differentiable at $x = x_0$ if and only if $\dfrac{F(x) - F(x_0^+)}{x - x_0}$ approaches a limit as $x \to x_0$ from the right. One denotes this one-sided limit, if it exists, by $\dfrac{d^+F(x_0)}{dx}$ and calls it the right-quasi-derivative of $F(x)$ at $x = x_0$. $F(x)$ is said to be left-quasi-differentiable at $x = x_0$ if and only if $\dfrac{F(x) - F(x_0^-)}{x - x_0}$ approaches a limit as $x \to x_0$ from the left. One denotes this one-sided limit, if it exists, by $\dfrac{d^-F(x_0)}{dx}$ and calls it the left-quasi-derivative of $F(x)$ at $x = x_0$.

Example 1. Consider the function defined for all x by

$$F(x) = \begin{cases} |x|, & -\pi \leq x \leq \pi \\ F(x + 2\pi). \end{cases}$$

Then $\quad \dfrac{d^+F(0)}{dx} = 1, \quad \dfrac{d^-F(0)}{dx} = -1, \quad \dfrac{d^+F(1)}{dx} = 1, \quad \dfrac{d^-F(1)}{dx} = 1,$

$$\frac{d^+F(-1)}{dx} = -1, \quad \frac{d^-F(-1)}{dx} = -1.$$

Example 2. Consider the function $F(x)$ defined on $-\pi \leq x \leq \pi$ by

$$F(x) = \begin{cases} 1, & 0 < x \leq \pi \\ 0, & x = 0 \\ -1, & -\pi \leq x < 0. \end{cases}$$

Then $\dfrac{d^-F(\pi)}{dx} = 0$, $\dfrac{d^+F(\pi)}{dx}$ is not defined, $\dfrac{d^+F(1)}{dx} = 0$, $\dfrac{d^-F(1)}{dx} = 0$,

$$\frac{d^+F(0)}{dx} = 0, \quad \frac{d^-F(0)}{dx} = 0.$$

Definition 2.6. A function $F(x)$ is said to be quasi-differentiable at a point $x = a$ if and only if it is both left- and right-quasi-differentiable at the point.

Example. Consider the function $F(x)$ defined on $-\pi \leq x \leq \pi$ by

$$F(x) = \begin{cases} 1, & 0 < x \leq \pi \\ 0, & x = 0 \\ -1, & -\pi \leq x < 0. \end{cases}$$

Then since $\dfrac{d^+F(0)}{dx}$ and $\dfrac{d^-F(0)}{dx}$ exist, $F(x)$ is quasi-differentiable at $x = 0$.
Note, of course, that even though, in this case, the left- and right-quasi-derivatives exist and both equal zero, the function is still not differentiable at $x = 0$.

Definition 2.7. Let a, b be two real numbers which satisfy $a < b$. Let M be the set of all real numbers x which satisfy $a \leq x \leq b$. A function $F(x)$ defined on M is said to be quasi-differentiable on M if and only if $F(x)$ is:
  (a) right-quasi-differentiable at $x = a$,
  (b) left-quasi-differentiable at $x = b$, and
  (c) quasi-differentiable for all x which satisfy $a < x < b$.

Note, with respect to Definition 2.7, that if $F(x)$ is differentiable at $x = x_0$, where $a < x_0 < b$, then it is of course quasi-differentiable at $x = x_0$ while, as indicated in the last example, the converse need not be valid at all.

Suppose now that, for $a < b$, $F(x)$ is continuous and $F'(x)$ is sectionally continuous on $a \leq x \leq b$. Moreover, let $F'(x)$ be continuous on $a < x < b$. It follows from Definition 2.4 that $F'(a^+)$, $F'(b^-)$ exist. One might reasonably ask whether or not the quasi-derivatives $\dfrac{d^+F(a)}{dx}$, $\dfrac{d^-F(b)}{dx}$ exist and whether or not $F'(a^+) = \dfrac{d^+F(a)}{dx}$,

$F'(b^-) = \dfrac{d^-F(b)}{dx}$. The following lemma answers this question in the affirmative.

Lemma 2.4. Let a, b be two real numbers such that $a < b$. Let M be the set of all real numbers x for which $a \leq x \leq b$. Let $F(x)$ be continuous on M and let $F'(x)$ be sectionally continuous on M. Moreover, assume $F'(x)$ is continuous on $a < x < b$. Then $\dfrac{d^+F(a)}{dx}$, $\dfrac{d^-F(b)}{dx}$ exist and

$$F'(a^+) = \frac{d^+F(a)}{dx} \; ; \qquad F'(b^-) = \frac{d^-F(b)}{dx} \; .$$

Proof. Let x_0 be a real number such that $a < x_0 < b$. On $a \leq x \leq x_0$, $F(x)$ satisfies the mean-value theorem so that

$$\frac{F(x_0) - F(a)}{x_0 - a} = F'(\xi), \qquad a < \xi < x_0. \tag{2.20}$$

Taking the right-hand limit of both sides of (2.20) as x_0 goes to a yields

$$\frac{d^+F(a)}{dx} = F'(a^+). \tag{2.21}$$

Since $F'(a^+)$ exists by Definition 2.4, then (2.21) implies $\dfrac{d^+F(a)}{dx}$ exists and equals $F'(a^+)$.

A similar argument reveals that $\dfrac{d^-F(b)}{dx}$ exists and that

$$\frac{d^-F(b)}{dx} = F'(b^-), \tag{2.22}$$

which completes the proof.

Lemma 2.5. Let a, b be two real numbers for which $a < b$. Let M be the set of all real numbers x for which $a \leq x \leq b$. Let $F(x)$ be continuous on M and let $F'(x)$ be sectionally continuous on M. Then $F(x)$ is quasi-differentiable on M.

Proof. By considering the finite number of subsets of M on which $F'(x)$ is continuous and by applying Lemma 2.4, the proof follows readily.

Sufficient conditions will now be established which ensure that the Fourier series associated with a function $F(x)$ actually converges to and represents that function for all real x.

Theorem 2.1. Let M be the set of all real numbers x for which $-\pi \leq x \leq \pi$. Let $F(x)$ be defined for all real x, have period 2π, and be sectionally continuous on M. Then at every point where $F(x)$ is *quasi-differentiable*, the Fourier series associated with $F(x)$ converges to $\dfrac{F(x^+) + F(x^-)}{2}$; that is, the Fourier series associated with $F(x)$ converges to and represents $F(x)$ at each point where $F(x)$ is quasi-differentiable and satisfies

$$F(x) = \frac{F(x^+) + F(x^-)}{2}. \tag{2.23}$$

Proof. Let

$$S_n(x) = \frac{a_0}{2} + \sum_{k=1}^{n} (a_k \cos kx + b_k \sin kx), \tag{2.24}$$

where $a_0, a_k, b_k;\ k = 1, 2, \ldots, n$ are given by (2.13) and (2.14). Then, by the elementary properties of integration, one has

$$S_n(x) = \frac{1}{2\pi} \int_{-\pi}^{\pi} F(t)\, dt + \sum_{k=1}^{n} \left\{ \left[\frac{1}{\pi} \int_{-\pi}^{\pi} F(t) \cos kt\, dt\right] \cos kx \right.$$

$$\left. + \left[\frac{1}{\pi} \int_{-\pi}^{\pi} F(t) \sin kt\, dt\right] \sin kx \right\}$$

$$= \frac{1}{\pi} \int_{-\pi}^{\pi} F(t) \left\{\frac{1}{2} + \sum_{k=1}^{n} (\cos kt \cos kx + \sin kt \sin kx)\right\} dt$$

$$= \frac{1}{\pi} \int_{-\pi}^{\pi} F(t) \left\{\frac{1}{2} + \sum_{k=1}^{n} \cos k(t - x)\right\} dt.$$

By (2.1) and (2.2), it follows that

$$S_n(x) = \frac{1}{\pi} \int_{-\pi}^{\pi} F(t) \frac{\sin \{(n + \frac{1}{2})(t - x)\}}{2 \sin [(t - x)/2]}\, dt.$$

The change of variables $v = t - x$ then implies

$$S_n(x) = \frac{1}{2\pi} \int_{-\pi-x}^{\pi-x} F(x + v) \frac{\sin (n + \frac{1}{2})v}{\sin (v/2)}\, dv.$$

By the periodicity of the integrand, it follows that

$$S_n(x) = \frac{1}{2\pi} \int_{-\pi}^{\pi} F(x + v) \frac{\sin (n + \frac{1}{2})v}{\sin (v/2)}\, dv.$$

For notational purposes, let $m = n + \frac{1}{2}$, so that

$$S_n(x) = \frac{1}{2\pi} \int_{-\pi}^{\pi} F(x + v) \frac{\sin mv}{\sin (v/2)}\, dv$$

$$= \frac{1}{2\pi} \int_{-\pi}^{0} F(x + v) \frac{\sin mv}{\sin (v/2)}\, dv + \frac{1}{2\pi} \int_{0}^{\pi} F(x + v) \frac{\sin mv}{\sin (v/2)}\, dv. \tag{2.25}$$

In an attempt to further simplify $S_n(x)$, it is convenient to introduce the function $s(v)$ which is defined at all points x where $F(x)$ is quasi-differentiable by

$$s(v) = \begin{cases} \dfrac{F(x+v) - F(x^+)}{2 \sin (v/2)}, & -\pi \leq v < 0, 0 < v \leq \pi \\[3mm] \dfrac{d^+ F(x)}{dx}, & v = 0 \end{cases} \qquad (2.26)$$

Note that, for fixed x, $s(v)$ is sectionally continuous in v on $0 \leq v \leq \pi$, for it is obviously so on $0 < v \leq \pi$, while at $v = 0$ the assertion follows from the existence of the right-quasi-derivative, since

$$\lim_{v \to 0^+} \frac{F(x+v) - F(x^+)}{2 \sin (v/2)} = \lim_{v \to 0^+} \left[\frac{F(x+v) - F(x^+)}{2 \sin (v/2)} \right] \lim_{v \to 0^+} \left[\frac{2 \sin (v/2)}{v} \right]$$

$$= \lim_{v \to 0^+} \frac{F(x+v) - F(x^+)}{v}$$

$$= \frac{d^+ F(x)}{dx}.$$

Hence,

$$\frac{1}{2\pi} \int_0^\pi F(x+v) \frac{\sin mv}{\sin (v/2)} \, dv = \frac{1}{\pi} \int_0^\pi s(v) \sin mv \, dv$$

$$+ \frac{1}{2\pi} \int_0^\pi F(x^+) \frac{\sin mv}{\sin (v/2)} \, dv. \qquad (2.27)$$

As $n \to \infty$, then $m \to \infty$, so that (2.27) and (2.15) imply

$$\lim_{n \to \infty} \frac{1}{2\pi} \int_0^\pi F(x+v) \frac{\sin mv}{\sin (v/2)} dv = \lim_{m \to \infty} \frac{1}{2\pi} \int_0^\pi F(x^+) \frac{\sin mv}{\sin (v/2)} \, dv. \qquad (2.28)$$

However, $F(x^+)$ is independent of v and, from (2.1) and (2.2), it follows immediately that

$$\int_0^\pi \frac{\sin mv}{2 \sin (v/2)} \, dv = \frac{\pi}{2},$$

so that (2.28) reduces to

$$\lim_{n \to \infty} \left[\frac{1}{2\pi} \int_0^\pi F(x+v) \frac{\sin mv}{\sin (v/2)} \, dv \right] = \frac{F(x^+)}{2}. \qquad (2.29)$$

In a similar fashion,

$$\lim_{n \to \infty} \left[\frac{1}{2\pi} \int_{-\pi}^0 F(x+v) \frac{\sin mv}{\sin (v/2)} \, dv \right] = \frac{F(x^-)}{2}. \qquad (2.30)$$

Thus, (2.29) and (2.30) imply

$$\lim_{n\to\infty}\left[\frac{1}{2\pi}\int_{-\pi}^{\pi}F(x+v)\frac{\sin mv}{\sin(v/2)}\,dv\right]=\frac{F(x^+)+F(x^-)}{2}. \quad (2.31)$$

Finally, by (2.24), (2.25), and (2.31), it follows that

$$\lim_{n\to\infty}S_n(x)=\lim_{n\to\infty}\left[\frac{1}{2\pi}\int_{-\pi}^{\pi}F(x+v)\frac{\sin mx}{\sin(v/2)}\,dv\right]$$

$$=\frac{F(x^+)+F(x^-)}{2},$$

and the theorem is proved.

Example 1. Consider the periodic function $F(x)$ defined for all x by

$$F(x)=\begin{cases}|x|, & -\pi\le x\le\pi\\ F(x+2\pi).\end{cases}$$

It was shown previously that the Fourier series associated with $F(x)$ is

$$\frac{\pi}{2}+\sum_{n=1}^{\infty}\left\{\frac{2}{\pi n^2}\left[(\cos n\pi)-1\right]\cos nx\right\}. \quad (2.32)$$

Since $F(x+2\pi)=F(x)$ and

(a) $F(x)$ is continuous and hence sectionally continuous on $-\pi\le x\le\pi$,

(b) $F(x)=\dfrac{F(x^+)+F(x^-)}{2}$ for all x since $F(x)$ is continous for all x, and

(c) $F(x)$ is quasi-differentiable for all x,

it follows that for all x the Fourier series (2.32) converges and represents the function under consideration, that is,

$$F(x)\equiv\frac{\pi}{2}+\sum_{n=1}^{\infty}\left\{\frac{2}{\pi n^2}\left[(\cos n\pi)-1\right]\cos nx\right\}.$$

Example 2. Consider the periodic function defined for all x by

$$F(x)=\begin{cases}x, & -\pi<x<\pi\\ 0, & x=\pi,-\pi\\ F(x+2\pi).\end{cases}$$

Since $F(x+2\pi)=F(x)$ and

(a) $F(x)$ is sectionally continuous on $-\pi\le x\le\pi$,

(b) $F(x)=\dfrac{F(x^+)+F(x^-)}{2}$ for all x, and

(c) $F(x)$ is quasi-differentiable for all x,

it follows that the Fourier series

$$F(x) = 2 \sum_{n=1}^{\infty} \frac{(-1)^{n+1}}{n} \sin nx$$

converges and represents $F(x)$ for all x.

Corollary. Let $F(x)$ be defined and continuous for all real x and have period 2π. Then if $F'(x)$ exists and is continuous on $-\pi \leq x \leq \pi$, the Fourier series associated with $F(x)$ converges to $F(x)$ for all real x.

2.2 Half-range Series

Often one must consider a function $f(x)$ which is defined only on $0 < x < \pi$. If one desires a Fourier series representation for $f(x)$ on $0 < x < \pi$, one could first seek a function $F(x)$ which has period 2π, which has a Fourier series representation, and which coincides with $f(x)$ on $0 < x < \pi$. Such a function $F(x)$, if it exists, is called a *periodic extension* of $f(x)$. The Fourier series representation for $F(x)$ would then be a representation for $f(x)$ on $0 < x < \pi$. After some thought, it is a relatively simple matter in many cases to find a variety of such periodic extensions, only two of which will now be explored.

Definition 2.8. A function $F(x)$, defined for all real x, is said to be an even function if and only if $F(-x) \equiv F(x)$.

Example 1. $F(x)$, defined for all real x by $F(x) = \cos x$, is an even function.

Example 2. $F(x)$, defined for all real x by

$$F(x) = \begin{cases} |x|, & -\pi \leq x \leq \pi \\ F(x + 2\pi), & \end{cases}$$

is an even function.

Theorem 2.2. Let M be the set of all real numbers x. Let $F(x)$ be defined on M, have period 2π, and have a convergent Fourier series representation on M. Then if $F(x)$ is an even function, the Fourier series representation of $F(x)$ is of the form

$$F(x) = \frac{a_0}{2} + \sum_{n=1}^{\infty} a_n \cos nx. \tag{2.33}$$

Proof.   In order to establish (2.33), one need only show that all the Fourier coefficients b_n; $n = 1, 2, \ldots$, are zero.   Hence, consider

$$b_n = \frac{1}{\pi} \int_{-\pi}^{\pi} F(x) \sin nx \, dx = \frac{1}{\pi} \int_{-\pi}^{0} F(x) \sin nx \, dx + \frac{1}{\pi} \int_{0}^{\pi} F(x) \sin nx \, dx.$$

$$(2.34)$$

By the elementary rules of integration and the fact that $F(x)$ is even, the transformation $u = -x$ yields

$$\int_{-\pi}^{0} F(x) \sin nx \, dx = -\int_{\pi}^{0} F(-u) \sin (-nu) \, du$$

$$= -\int_{0}^{\pi} F(u) \sin nu \, du.$$

However, since u is only a dummy variable, this latter result may be restated as

$$\int_{-\pi}^{0} F(x) \sin nx \, dx = -\int_{0}^{\pi} F(x) \sin nx \, dx. \qquad (2.35)$$

Substitution of (2.35) into (2.34) implies $b_n = 0$; $n = 1, 2, \ldots$, and the theorem is proved.

Example.   The function $F(x)$ defined for all real x by

$$F(x) = \begin{cases} |x|, & -\pi \leq x \leq \pi \\ F(x + 2\pi) \end{cases}$$

is even and has Fourier series representation

$$F(x) \equiv \frac{\pi}{2} + \sum_{n=1}^{\infty} \left\{ \frac{2}{\pi n^2} [(\cos n\pi) - 1] \cos nx \right\}.$$

Definition 2.9.   A function $F(x)$, defined for all real x, is said to be an odd function if and only if $F(-x) \equiv -F(x)$.

Example 1. $F(x)$, defined for all real x by $F(x) = \sin x$, is an odd function.

Example 2. $F(x)$, defined for all real x by

$$F(x) = \begin{cases} x, & -\pi < x < \pi \\ 0, & x = \pi, -\pi \\ F(x + 2\pi), \end{cases}$$

is an odd function.

Theorem 2.3.   Let M be the set of all real numbers x.   Let $F(x)$ be defined on M, have period 2π, and have a convergent Fourier series representation on M. Then if $F(x)$ is an odd function, the Fourier series representation of $F(x)$ is of the form

$$F(x) = \sum_{n=1}^{\infty} b_n \sin nx. \qquad (2.36)$$

Proof. In order to establish (2.36), one need only show that all the Fourier coefficients a_n $(n = 0, 1, 2, \ldots)$ are zero. The proof is completely analogous, then, to that of Theorem 2.2.

Example. The function $F(x)$ defined for all real x by

$$F(x) = \begin{cases} x, & -\pi < x < \pi \\ 0, & x = \pi, -\pi \\ F(x + 2\pi) \end{cases}$$

is odd and has the Fourier series representation

$$F(x) = 2 \sum_{n=1}^{\infty} \frac{(-1)^{n+1}}{n} \sin nx.$$

Definition 2.10. Let $f(x)$ be defined only for $0 < x < \pi$. Then the function $F(x)$ defined for all real x by

$$F(x) = \begin{cases} f(x), & 0 < x < \pi \\ f(-x), & -\pi < x < 0 \\ f(0^+), & x = 0 \\ f(\pi^-), & x = \pi, -\pi \\ F(x + 2\pi), \end{cases}$$

provided all the indicated limits exist, is called the even periodic extension of $f(x)$.

Example. Let $f(x)$ be defined on $0 < x < \pi$ by $f(x) = 1 + x$. Then the even periodic extension of $f(x)$ is the function $F(x)$ which is defined for all real x by

$$F(x) = \begin{cases} 1 + x, & 0 \le x \le \pi \\ 1 - x, & -\pi \le x \le 0 \\ F(x + 2\pi). \end{cases}$$

Note, of course, that the even periodic extension, as defined above, is an even function.

Definition 2.11. Let $f(x)$ be defined only for $0 < x < \pi$. Then the function $F(x)$ defined for all real x by

$$F(x) = \begin{cases} f(x), & 0 < x < \pi \\ -f(-x), & -\pi < x < 0 \\ 0, & x = 0, \pi, -\pi \\ F(x + 2\pi) \end{cases}$$

is called the odd periodic extension of $f(x)$.

Example. Let $f(x)$ be defined on $0 < x < \pi$ by $f(x) = 1 + x$. Then the odd periodic extension of $f(x)$ is the function $F(x)$ which is defined for all real x by

$$F(x) = \begin{cases} 1 + x, & 0 < x < \pi \\ -1 + x, & -\pi < x < 0 \\ 0, & x = 0, \pi, -\pi \\ F(x + 2\pi). \end{cases}$$

Note, of course, that an odd periodic extension, as defined above, is an odd function.

Definition 2.12. Let $f(x)$ be defined only for $0 < x < \pi$. If the even periodic extension $F(x)$ has a Fourier series representation which is valid for all real x, and must, by Theorem 2.2, be of the form

$$\frac{a_0}{2} + \sum_{n=1}^{\infty} a_n \cos nx, \tag{2.37}$$

then series (2.37) is called the half-range Fourier cosine series for $f(x)$.

Example. Let $f(x)$ be defined on $0 < x < \pi$ by $f(x) = 1 + x$. Then its half-range Fourier cosine series is

$$\frac{\pi + 2}{2} + \sum_{n=1}^{\infty} \frac{2}{\pi n^2} [(\cos n\pi) - 1] \cos nx.$$

Definition 2.13. Let $f(x)$ be defined only for $0 < x < \pi$. If the odd periodic extension $F(x)$ has a Fourier series representation which is valid for all real x, and must, by Theorem 2.3, be of the form

$$\sum_{n=1}^{\infty} b_n \sin nx, \tag{2.38}$$

then series (2.38) is called the half-range Fourier sine series for $f(x)$.

Example. Let $f(x)$ be defined on $0 < x < \pi$ by $f(x) = 1 + x$. Then its half-range Fourier sine series is

$$\sum_{n=1}^{\infty} \frac{2}{\pi n} (1 - \cos n\pi - \pi \cos n\pi) \sin nx.$$

2.3 Change of Interval

If a function $F(x)$, defined for all real x, has period $2L$, then the change of variable $x' = \pi x / L$ yields the following more general form of Theorem 2.1.

Theorem 2.4. If L is a positive, real number, let M be the set of all real numbers x for which $-L \leq x \leq L$. Let $F(x)$ be defined for all real x, have period $2L$, and be sectionally continuous on M. Then, at every point where $F(x)$ is quasi-differentiable, the Fourier series

$$\frac{a_0}{2} + \sum_{n=1}^{\infty} \left(a_n \cos \frac{n\pi x}{L} + b_n \sin \frac{n\pi x}{L} \right),$$

where

$$a_n = \frac{1}{L} \int_{-L}^{L} F(x) \cos \frac{n\pi x}{L} \, dx, \qquad n = 0, 1, 2, \ldots,$$

$$b_n = \frac{1}{L} \int_{-L}^{L} F(x) \sin \frac{n\pi x}{L} \, dx, \qquad n = 1, 2, \ldots,$$

converges to

$$\frac{F(x^+) + F(x^-)}{2}.$$

Example. The function $F(x)$ with period 4, defined for all real x by

$$F(x) = \begin{cases} 3, & 0 < x < 2 \\ 0, & -2 < x < 0 \\ \frac{3}{2}, & x = 2, 0, -2 \\ F(x + 4), & \end{cases}$$

satisfies $\dfrac{F(x^+) + F(x^-)}{2} = F(x)$ for all real x and is quasi-differentiable for all real x. Thus, for all real x,

$$F(x) \equiv \frac{3}{2} + \sum_{n=1}^{\infty} \frac{3}{n\pi} (1 - \cos n\pi) \sin \frac{n\pi x}{2}.$$

2.4 Differentiation of Fourier Series

Since Fourier series will be used to generate solutions of differential equations, it is reasonable to seek conditions which ensure that the Fourier series representation of a function is term-wise differentiable. Moreover, if for $a < b$ a Fourier series is given on $a \leq x \leq b$, later considerations will require differentiability only on $a < x < b$. The following theorem then will suffice for our purposes.

Theorem 2.5. Let M denote the set of real numbers x such that $-\pi \leq x \leq \pi$ while M_0 denotes the set of real numbers x such that $-\pi < x < \pi$. Let $F(x) \in C^1$ on M and assume $F(\pi) = F(-\pi)$.

Then, at each point of M_0 where $F''(x)$ exists, the Fourier series for $F(x)$,

$$\frac{a_0}{2} + \sum_{n=1}^{\infty} (a_n \cos nx + b_n \sin nx)$$

is differentiable and its derivative converges to $F'(x)$; that is, at each point of M_0 where $F''(x)$ exists, the series

$$\sum_{n=1}^{\infty} (-na_n \sin nx + nb_n \cos nx)$$

converges and

$$F'(x) = \sum_{n=1}^{\infty} (-na_n \sin nx + nb_n \cos nx). \qquad (2.39)$$

Proof. Since $F'(x)$ is continuous on M, by the corollary to Theorem 2.1 it follows that at each $x \in M$, the Fourier series for $F(x)$ converges and represents $F(x)$, that is, on M,

$$F(x) = \frac{a_0}{2} + \sum_{n=1}^{\infty} (a_n \cos nx + b_n \sin nx).$$

Again, since $F'(x)$ is continuous on M, it follows readily, with the aid of Theorem 2.1, that at each point $x \in M_0$ at which $F''(x)$ exists, one has

$$F'(x) = \frac{\bar{a}_0}{2} + \sum_{n=1}^{\infty} (\bar{a}_n \cos nx + \bar{b}_n \sin nx), \qquad (2.40)$$

where

$$\bar{a}_n = \frac{1}{\pi} \int_{-\pi}^{\pi} F'(x) \cos nx \, dx, \qquad \bar{b}_n = \frac{1}{\pi} \int_{-\pi}^{\pi} F'(x) \sin nx \, dx.$$

Integration by parts and use of the assumption that $F(\pi) = F(-\pi)$ yields

$$\bar{a}_n = \frac{1}{\pi} [F(x) \cos nx] \Big|_{-\pi}^{\pi} + \frac{n}{\pi} \int_{-\pi}^{\pi} F(x) \sin nx \, dx$$

$$= \frac{\cos n\pi}{\pi} [F(\pi) - F(-\pi)] + nb_n$$

$$= nb_n.$$

Similarly,

$$\bar{b}_n = -na_n.$$

Substitution of these values of $\bar{a}_n$, $\bar{b}_n$ into (2.40) yields (2.39) and the theorem is proved.

Example. Consider

$$F(x) = x^2 - \pi^2, \qquad -\pi \le x \le \pi.$$

Then
$$F'(x) = 2x, \qquad -\pi \le x \le \pi.$$
All conditions of Theorem 2.5 are satisfied and $F''(x)$ exists at each point of $-\pi < x < \pi$. Thus
$$x^2 - \pi^2 \equiv -\frac{2\pi^2}{3} + \sum_{n=1}^{\infty}\left[\frac{4}{n^2}\cos n\pi \cos nx\right], \qquad -\pi \le x \le \pi,$$
and, by termwise differentiation,
$$2x \equiv \sum_{n=1}^{\infty}\left(-\frac{4}{n}\cos n\pi \sin nx\right), \qquad -\pi < x < \pi.$$
Also, note that this last identity is actually *not* valid at $x = \pi, -\pi$. In order to establish a theorem analogous to Theorem 2.5 which would be valid at these points, one would have to take into consideration the result that a convergent Fourier series converges according to (2.23). This, however, will not be necessary for our purposes.

2.5   Absolute and Uniform Convergence

It will be convenient when constructing solutions of partial differential equations by means of Fourier series to have sufficient conditions for the absolute and uniform convergence of such series.

Lemma 2.6.   Let $a_1, a_2, \ldots, a_n$ and $b_1, b_2, \ldots, b_n$ be any real numbers.   Then the Cauchy-Schwarz inequality is valid, that is,

$$\left[\sum_{i=1}^{n}(a_i b_i)\right]^2 \le \left(\sum_{i=1}^{n}a_i^2\right)\left(\sum_{i=1}^{n}b_i^2\right). \qquad (2.41)$$

Proof.

$$\left(\sum_{i=1}^{n}a_i^2\right)\left(\sum_{i=1}^{n}b_i^2\right) - \frac{1}{2}\sum_{i=1}^{n}\sum_{j=1}^{n}(a_i b_j - b_i a_j)^2$$

$$\equiv \left(\sum_{i=1}^{n}a_i^2\right)\left(\sum_{i=1}^{n}b_i^2\right) - \frac{1}{2}\left(\sum_{i=1}^{n}\sum_{j=1}^{n}a_i^2 b_j^2\right) - \frac{1}{2}\left(\sum_{i=1}^{n}\sum_{j=1}^{n}a_j^2 b_i^2\right)$$
$$+ \sum_{i=1}^{n}\sum_{j=1}^{n}(a_i b_j a_j b_i)$$

$$\equiv \left(\sum_{i=1}^{n}a_i^2\right)\left(\sum_{i=1}^{n}b_i^2\right) - \frac{1}{2}\left(\sum_{i=1}^{n}a_i^2\sum_{i=1}^{n}b_i^2\right) - \frac{1}{2}\left(\sum_{i=1}^{n}a_i^2\sum_{i=1}^{n}b_i^2\right)$$
$$+ \sum_{i=1}^{n}(a_i b_i)\sum_{j=1}^{n}(a_j b_j)$$

$$\equiv \left[\sum_{i=1}^{n}(a_i b_i)\right]\left[\sum_{i=1}^{n}(a_i b_i)\right]$$

$$\equiv \left[\sum_{i=1}^{n}(a_i b_i)\right]^2.$$

Hence,

$$\left[\sum_{i=1}^{n}(a_i b_i)\right]^2 \equiv \left(\sum_{i=1}^{n}a_i^2\right)\left(\sum_{i=1}^{n}b_i^2\right) - \frac{1}{2}\sum_{i=1}^{n}\sum_{j=1}^{n}(a_i b_j - b_i a_j)^2,$$

which implies (2.41), and the lemma is proved.

Note that the Cauchy-Schwarz inequality is also called Buniakovsky's inequality.

Lemma 2.7. Let M be the set of all real x such that $-\pi \leq x \leq \pi$. Let $F(x) \in C^1$ on M, so that $F(x)$ is represented everywhere on M by its Fourier series

$$F(x) = \frac{a_0}{2} + \sum_{n=1}^{\infty}(a_n \cos nx + b_n \sin nx).$$

Then, for all positive, integral values of k, the Bessel inequality, namely,

$$\tfrac{1}{2}a_0^2 + \sum_{n=1}^{k}(a_n^2 + b_n^2) \leq \frac{1}{\pi}\int_{-\pi}^{\pi}[F(x)]^2\,dx \tag{2.42}$$

is valid.

Proof. Noting that

$$\int_{-\pi}^{\pi}\left[F(x) - \frac{a_0}{2} - \sum_{n=1}^{k}(a_n \cos nx + b_n \sin nx)\right]^2 dx \geq 0$$

and, with the aid of (2.9), that

$$\int_{-\pi}^{\pi}\left[F(x) - \frac{a_0}{2} - \sum_{n=1}^{k}(a_n \cos nx + b_n \sin nx)\right]^2 dx$$

$$= \int_{-\pi}^{\pi}\left\{[F(x)]^2 + \left(\frac{a_0}{2}\right)^2 + \sum_{n=1}^{k}(a_n^2 \cos^2 nx + b_n^2 \sin^2 nx) - a_0 F(x)\right.$$

$$\left. - 2\sum_{n=1}^{k}(a_n F(x)\cos nx + b_n F(x)\sin nx)\right\}dx$$

$$= \int_{-\pi}^{\pi}[F(x)]^2\,dx - \pi\left[\frac{a_0^2}{2} + \sum_{n=1}^{k}(a_n^2 + b_n^2)\right],$$

leads immediately to (2.42) and the lemma is proved.

Theorem 2.6. Denote the set of all real x such that $-\pi \leq x \leq \pi$ by M and the set of all real x such that $-\pi < x < \pi$ by M_0. Let $F(x) \in C^2$ on M. If, in addition, $F(\pi) = F(-\pi)$, then the Fourier series for $F(x)$ on M converges absolutely and uniformly on M_0.

Proof. By means of the corollary to Theorem 2.1, on M

$$F(x) = \frac{a_0}{2} + \sum_{n=1}^{\infty}(a_n \cos nx + b_n \sin nx), \tag{2.43}$$

and, by Theorem 2.5, on M_0

$$F'(x) = \sum_{n=1}^{\infty} (-na_n \sin nx + nb_n \cos nx). \qquad (2.44)$$

Application of (2.42) to (2.44) yields

$$\sum_{n=1}^{k} [n^2(a_n^2 + b_n^2)] \le \frac{1}{\pi} \int_{-\pi}^{\pi} [F'(x)]^2 \, dx. \qquad (2.45)$$

Since the Cauchy-Schwarz inequality implies

$$(a_n \cos nx + b_n \sin nx)^2 \le (a_n^2 + b_n^2)(\cos^2 nx + \sin^2 nx) = (a_n^2 + b_n^2),$$

so that

$$|a_n \cos nx + b_n \sin nx| \le \sqrt{a_n^2 + b_n^2},$$

then, for $m > k$,

$$\sum_{n=k+1}^{m} |a_n \cos nx + b_n \sin nx| \le \sum_{n=k+1}^{m} \sqrt{a_n^2 + b_n^2}. \qquad (2.46)$$

However, with the aid of the Cauchy-Schwarz inequality, it readily follows that

$$\sum_{n=k+1}^{m} \sqrt{a_n^2 + b_n^2} = \sum_{n=k+1}^{m} \left[\frac{1}{n} (n\sqrt{a_n^2 + b_n^2}) \right]$$

$$\le \sqrt{\left[\sum_{n=k+1}^{m} \frac{1}{n^2} \right] \left[\sum_{n=k+1}^{m} n^2(a_n^2 + b_n^2) \right]}. \qquad (2.47)$$

Setting

$$N^2 = \frac{1}{\pi} \int_{-\pi}^{\pi} [F'(x)]^2 \, dx,$$

then (2.45) to (2.47) imply that for $x \in M_0$

$$\sum_{n=k+1}^{m} |a_n \cos nx + b_n \sin nx| \le |N| \sqrt{\sum_{n=k+1}^{m} \frac{1}{n^2}}.$$

However, since $\sum_{1}^{\infty} \frac{1}{n^2}$ converges and is independent of x, it follows that by choosing k sufficiently large, then, independently of m and $x \in M_0$,

$$\sum_{n=k+1}^{m} |a_n \cos nx + b_n \sin nx|$$

can be made arbitrarily small, from which it follows that (2.43) converges absolutely and uniformly on M_0.

Corollary 1. Under the hypotheses of Theorem 2.6, each of the series $\sum\limits_{n=1}^{\infty} \sqrt{a_n^2 + b_n^2}, \sum\limits_{n=1}^{\infty} |a_n|, \sum\limits_{n=1}^{\infty} |b_n|$ converges.

Proof. The corollary follows from the proof of Theorem 2.6.

Corollary 2. Under the hypotheses of Theorem 2.6, the given Fourier series converges absolutely and uniformly on M.

Proof. The absolute convergence follows from Corollary 1 and the uniform convergence follows from the assumption that $F(x)$ is continuous on M, from the assumption that $F(\pi) = F(-\pi)$, from Theorem 2.6, and from the definition of uniform convergence.

Example. Since the function

$$F(x) = x^2 - \pi^2, \qquad -\pi \leq x \leq \pi,$$

satisfies the conditions of Theorem 2.6, then it readily follows that the series

$$-\frac{2\pi^2}{3} + \sum_{n=1}^{\infty} \left(\frac{4}{n^2} \cos n\pi \cos nx \right)$$

converges absolutely and uniformly on $-\pi \leq x \leq \pi$.

2.6 A Weierstrass Approximation Theorem

Definition 2.14. Let M be the set of all real numbers and let $F(x)$ be continuous on M and have period 2π. Suppose that on $-\pi \leq x \leq \pi$, there exists a finite number of points $-\pi = x_0 < x_1 < x_2 < x_3 < \cdots < x_{m-1} < x_m = \pi$ such that $F(x)$ is linear on $x_{j-1} \leq x \leq x_j, j = 1, 2, 3, \ldots, m$. Then $F(x)$ is called a *broken-line function*.

Example. The function $F(x)$, defined for all real x by

$$F(x) = \begin{cases} 7 + \dfrac{5(x + 1)}{\pi - 1}, & -\pi \leq x \leq -1 \\[2mm] 7 - \dfrac{8(x + 1)}{3}, & -1 \leq x \leq 2 \\[2mm] -1 + \dfrac{3(x - 2)}{\pi - 2}, & 2 \leq x \leq \pi \\[2mm] F(x + 2\pi), & \end{cases}$$

and exhibited in Diagram 2.1, is a broken-line function.

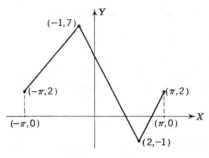

DIAGRAM 2.1

Lemma 2.8. The Fourier series for a broken-line function $F(x)$ converges uniformly to the function for all values of x.

Proof. Using the notation of Definition 2.14, for $x_{j-1} < x < x_j$, let $\alpha_j = F'(x)$, $j = 1, 2, \ldots, m$ and let $\alpha = \max\limits_{j} |\alpha_j|$. Then, for $k = 1, 2, \ldots, m$,

$$\int_{x_{j-1}}^{x_j} F(x) \cos kx \, dx = \frac{1}{k} F(x) \sin kx \Big|_{x_{j-1}}^{x_j} - \frac{1}{k} \int_{x_{j-1}}^{x_j} \alpha_j \sin kx \, dx$$

$$= \frac{1}{k} \left[F(x_j) \sin kx_j - F(x_{j-1}) \sin kx_{j-1} \right]$$

$$+ \frac{\alpha_j}{k^2} (\cos kx_j - \cos kx_{j-1}).$$

Moreover,

$$\sum_{j=1}^{m} \left[F(x_j) \sin kx_j - F(x_{j-1}) \sin kx_{j-1} \right]$$
$$= F(\pi) \sin kx - F(-\pi) \sin (-k\pi) = 0,$$

while

$$\left| \frac{\alpha_j}{k^2} (\cos kx_j - \cos kx_{j-1}) \right| \leq \frac{2\alpha}{k^2}$$

and

$$\left| \sum_{j=1}^{m} \frac{\alpha_j}{k^2} (\cos kx_j - \cos kx_{j-1}) \right| \leq \frac{2\alpha m}{k^2}.$$

Consequently,

$$|a_k| = \frac{1}{\pi} \left| \sum_{j=1}^{m} \int_{x_{j-1}}^{x_j} F(x) \cos kx \, dx \right| \leq \frac{2\alpha m}{\pi k^2}, \qquad k = 1, 2, \ldots.$$

Similarly,

$$|b_k| \leq \frac{2\alpha m}{\pi k^2}, \qquad k = 1, 2, \ldots,$$

so that the Fourier coefficients of $F(x)$ have the property that

$$|a_k| \le \frac{K}{k^2}, \qquad |b_k| \le \frac{K}{k^2}, k = 1, 2, \ldots, \qquad (2.48)$$

where K is a constant which is independent of k.

If $S_n(x)$ is the nth partial sum of the Fourier series for $F(x)$, then

$$F(x) - S_n(x) = \sum_{k=n+1}^{\infty} (a_k \cos kx + b_k \sin kx), \qquad n \ge 1. \qquad (2.49)$$

However, from (2.48) and the fundamental laws of inequalities, (2.49) yields

$$\begin{aligned}
|F(x) - S_n(x)| &= \left| \sum_{k=n+1}^{\infty} (a_k \cos kx + b_k \sin kx) \right| \\
&\le \sum_{k=n+1}^{\infty} (|a_k \cos kx| + |b_k \sin kx|) \\
&\le \sum_{k=n+1}^{\infty} \frac{2K}{k^2} \\
&= 2K \sum_{k=n+1}^{\infty} \frac{1}{k^2}.
\end{aligned}$$

However, since for $k \ge 1$, $k - 1 \le h \le k$ implies $1/k^2 \le 1/h^2$, it follows that

$$\frac{1}{k^2} = \int_{k-1}^{k} \frac{dh}{k^2} \le \int_{k-1}^{k} \frac{dh}{h^2}$$

and

$$\sum_{k=n+1}^{\infty} \frac{1}{k^2} \le \int_{n}^{\infty} \frac{dh}{h^2} = \frac{1}{n},$$

so that

$$|F(x) - S_n(x)| \le \frac{2K}{n}. \qquad (2.50)$$

Since the right-hand member of (2.50) is independent of x, the lemma readily follows.

Note that broken-line functions do not, in general, satisfy the conditions of Theorem 2.6, so that neither that theorem nor its corollaries were applicable and Lemma 2.8 had to be proved independently.

Theorem 2.7. (Weierstrass). Let M be the set of all real numbers. If $F(x)$ is continuous on M and has period 2π, then it can be uniformly approximated by a finite trigonometric sum with any preassigned degree of accuracy.

Proof. Let ϵ be an arbitrary positive number. Let $F^*(x)$ be a broken-line function which is defined for all real x, has period 2π, and satisfies, for all x,

$$|F(x) - F^*(x)| < \frac{\epsilon}{2}. \qquad (2.51)$$

The existence of $F^*(x)$ follows directly from the uniform continuity of $F(x)$ on $-\pi \leq x \leq \pi$.

If $F^*(x)$ has the Fourier series representation

$$F^*(x) = \frac{a_0}{2} + \sum_{k=1}^{\infty} (a_k \cos kx + b_k \sin kx), \qquad (2.52)$$

then, by Lemma 2.8, the series (2.52) converges uniformly to $F^*(x)$, that is, if

$$S_n(x) = \frac{a_0}{2} + \sum_{k=1}^{n} (a_k \cos kx + b_k \sin kx),$$

then for given $\epsilon > 0$, there exists N which is independent of x such that

$$|F^*(x) - S_N(x)| < \frac{\epsilon}{2}.$$

Hence, fix an N for the ϵ given above. Then

$$\begin{aligned}
|F(x) - S_N(x)| &= |F(x) - F^*(x) + F^*(x) - S_N(x)| \\
&\leq |F(x) - F^*(x)| + |F^*(x) - S_N(x)| \\
&\leq \frac{\epsilon}{2} + \frac{\epsilon}{2} \\
&= \epsilon,
\end{aligned}$$

and the theorem is proved.

2.7 Abel's Test

If one must consider a series like

$$\sum_{n=1}^{\infty} [(e^{-n^2 y})(b_n \sin nx)],$$

where each term of the series is constructed by multiplying each term $b_n \sin nx$ of a Fourier sine series by the term $e^{-n^2 y}$, a convenient test for convergence is Abel's test, which will now be developed.

Definition 2.15. If c, d are two real constants with $c < d$, then a sequence of functions $g_n(y)$; $n = 1, 2, 3, \ldots$, defined on $c \leq y \leq d$, is said to be uniformly bounded on $c \leq y \leq d$ if and only if there exists a constant A such that, independently of n and for all y in $c \leq y \leq d$,

$$|g_n(y)| < A.$$

Example. If $0 \leq c \leq d$, then the sequence of functions $\{e^{-n^2y}\}$ is uniformly bounded on $c \leq y \leq d$ since, independently of n and for all nonnegative y, $|e^{-n^2y}| < 2$.

It is now convenient to introduce the concept of a sequence of functions $\{g_n(y)\}$ which is monotone with respect to n.

Definition 2.16. If c, d are two real constants with $c < d$, then a sequence of functions $\{g_n(y)\}$, $c \leq y \leq d$, is said to be *monotonic increasing with respect to n* on $c \leq y \leq d$ if and only if for all y in $c \leq y \leq d$

$$g_{n+1}(y) \geq g_n(y), \qquad n = 1, 2, 3, \ldots, \tag{2.53}$$

while it is said to be *strictly* monotonic increasing with respect to n on $c \leq y \leq d$ if and only if for all y in $c \leq y \leq d$

$$g_{n+1}(y) > g_n(y), \qquad n = 1, 2, 3, \ldots. \tag{2.54}$$

Definition 2.17. If c, d are two real constants with $c < d$, then the sequence of functions $\{g_n(y)\}$, $c \leq y \leq d$, is said to be *monotonic decreasing with respect to n* on $c \leq y \leq d$ if and only if for all y in $c \leq y \leq d$

$$g_{n+1}(y) \leq g_n(y), \qquad n = 1, 2, 3, \ldots, \tag{2.55}$$

while it is said to be *strictly* monotonic decreasing with respect to n on $c \leq y \leq d$ if and only if for all y in $c \leq y \leq d$

$$g_{n+1}(y) < g_n(y), \qquad n = 1, 2, 3, \ldots. \tag{2.56}$$

Example 1. On $0 \leq y \leq 2$, $\{e^{ny}\}$ is monotonic increasing with respect to n.

Example 2. On $1 \leq y \leq 2$, $\{e^{ny}\}$ is strictly monotonic increasing with respect to n.

Example 3. On $0 \leq y \leq 2$, $\{e^{-n^2y}\}$ is monotonic decreasing with respect to n.

Example 4. On $1 \leq y \leq 2$, $\{e^{-n^2y}\}$ is strictly monotonic decreasing with respect to n.

Theorem 2.8. Abel's Test. Let a, b, c, d be four real numbers, with $a < b$, $c < d$. Let $\bar{R}$ be a closed region of the xy plane whose projection on the x axis consists of all x which satisfy $a \leq x \leq b$

and whose projection on the y axis consists of all y which satisfy $c \leq y \leq d$. If

(a) the series $\sum\limits_{n=1}^{\infty} F_n(x)$ converges uniformly with respect to x on $a \leq x \leq b$, and

(b) the functions $G_n(y)$, $n = 1, 2, 3, \ldots$, are uniformly bounded and either monotonic decreasing or monotonic increasing with respect to n on $c \leq y \leq d$,

then the series

$$\sum_{n=1}^{\infty} F_n(x)G_n(y)$$

converges uniformly in the two variables x and y together on $\bar{R}$.

Proof. Suppose first $\{G_n(y)\}$ is monotonic decreasing with respect to n on $c \leq y \leq d$. Let

$$S_n(x,y) = \sum_{i=1}^{n} [F_i(x)G_i(y)]. \tag{2.57}$$

In order to establish the theorem, it is sufficient to show that for any given $\epsilon > 0$, there exists a positive integer N, which depends on ϵ but is independent of x and y, such that for all $n > N$, all $m = n + 1, n + 2, \ldots$, and all $(x,y) \in \bar{R}$, one has

$$|S_n(x,y) - S_m(x,y)| < \epsilon.$$

For this purpose, let

$$s_n(x,y) = \sum_{i=1}^{n} F_i(x). \tag{2.58}$$

Then for every pair of integers m, n, with $m > n$, one has

$$\begin{aligned}
S_m - S_n &= \sum_{i=n+1}^{m} F_i G_i \\
&= \sum_{i=n+1}^{m} [(s_i - s_{i-1})G_i] \\
&= \sum_{i=n+1}^{m} [(s_i - s_n + s_n - s_{i-1})G_i] \\
&= \sum_{i=n+1}^{m} [(s_i - s_n)G_i] - \sum_{i=n+2}^{m} [(s_{i-1} - s_n)G_i] \\
&= \sum_{i=n+1}^{m} [(s_i - s_n)G_i] - \sum_{i=n+1}^{m-1} [(s_i - s_n)G_{i+1}],
\end{aligned}$$

so that

$$S_m - S_n = \sum_{i=n+1}^{m-1} [(s_i - s_n)(G_i - G_{i+1})] + (s_m - s_n)G_m. \quad (2.59)$$

By assumption, each factor $(G_i - G_{i+1})$ in (2.59) is nonnegative and there exists a constant A such that, independently of y on $c \le y \le d$, $|G_m| < A$. Since the series $\sum_{n=1}^{\infty} F_n(x)$ is uniformly convergent, an integer N can be found for which

$$|s_{n+r} - s_n| < \frac{\epsilon}{3A},$$

where ϵ is any positive number, N is independent of x, n is any positive integer larger than N, and r is an arbitrary nonnegative integer. For this choice of N, (2.59) implies

$$|S_m - S_n| \le \frac{\epsilon}{3A} \left[\sum_{i=n+1}^{m-1} (G_i - G_{i+1}) + |G_m| \right]$$

$$= \frac{\epsilon}{3A} (G_{n+1} - G_m + |G_m|)$$

$$\le \frac{\epsilon}{3A} (|G_{n+1}| + |G_m| + |G_m|)$$

$$< \frac{\epsilon}{3A} (3A)$$

$$= \epsilon,$$

so that

$$|S_m - S_n| < \epsilon. \quad (2.60)$$

Since a similar argument is valid if $G_n(y)$ is monotonic increasing with respect to n, the theorem then readily follows from (2.60).

Example. Let $F(x)$ be defined for all real x by

$$F(x) = \frac{2}{\pi} \sum_{n=1}^{\infty} \left\{ \left[\int_0^\pi F(t) \sin nt \, dt \right] \sin nx \right\}, \quad (2.61)$$

where $F(x)$ satisfies the conditions of Theorem 2.6.   Then (2.61) converges uniformly on, say, $\pi/4 \le x \le \pi/2$.   Recall that $e^{-n^2 y}$, $n = 1, 2, \ldots$, is monotonic decreasing and uniformly bounded on $0 \le y \le 2$.   Then by Abel's test the series

$$\frac{2}{\pi} \sum_{n=1}^{\infty} \left\{ [e^{-n^2 y} \sin nx] \left[\int_0^\pi F(t) \sin nt \, dt \right] \right\} \quad (2.62)$$

converges absolutely and uniformly in the two variables x and y together on the closed region $\bar{R}$ which consists of all (x,y) whose coordinates satisfy $\pi/4 \leq x \leq \pi/2,\ 0 \leq y \leq 2$.

EXERCISES

2.1. Each of the following functions is defined for all real x. Graph each function:

(a) $F(x) = \displaystyle\sum_{n=1}^{2} \sin nx$

(b) $F(x) = \displaystyle\sum_{n=1}^{3} \cos nx$

(c) $F(x) = \displaystyle\sum_{n=1}^{2} (n \sin nx + n^2 \cos nx)$.

2.2. Graph each of the following functions. In each case find the Fourier series representation of the function and show that for all real x, the series converges and represents the function:

(a) $F(x) = \begin{cases} 2, & 0 < x < \pi \\ -1, & -\pi < x < 0 \\ \frac{1}{2}, & x = -\pi, 0, \pi \\ F(x + 2\pi). \end{cases}$ $\left(Ans.\ \ \dfrac{1}{2} + \dfrac{6}{\pi} \displaystyle\sum_{n=1}^{\infty} \dfrac{\sin (2n - 1)x}{2n - 1} \right)$

(b) $F(x) = \begin{cases} 0, & 0 \leq x \leq \pi \\ \sin x, & -\pi \leq x \leq 0 \\ F(x + 2\pi). \end{cases}$

(c) $F(x) = \begin{cases} x^2, & -\pi \leq x \leq \pi \\ F(x + 2\pi). \end{cases}$ $\left(Ans.\ \ \dfrac{\pi^2}{3} - 4 \displaystyle\sum_{n=1}^{\infty} \dfrac{(-1)^{n+1} \cos nx}{n^2} \right)$

(d) $F(x) = \begin{cases} 0, & 0 < x < \pi \\ \pi, & -\pi < x < 0 \\ \dfrac{\pi}{2}, & x = 0, \pi, -\pi \\ F(x + 2\pi). \end{cases}$

(e) $F(x) = \begin{cases} x \cos x, & -\pi < x < \pi \\ 0, & x = \pi, -\pi \\ F(x + 2\pi). \end{cases}$ $\left(Ans.\ \ -\tfrac{1}{2} \sin x + 2 \displaystyle\sum_{n=2}^{\infty} \dfrac{(-1)^n n}{n^2 - 1} \sin nx \right)$

(f) $F(x) = \begin{cases} -1, & -\pi < x < 0 \\ 1, & 0 < x < \pi \\ 0, & x = 0, \pi, -\pi \\ F(x + 2\pi). \end{cases}$

(g) $F(x) = \begin{cases} x, & 0 < x < \pi \\ -\pi, & -\pi < x < 0 \\ 0, & x = \pi, -\pi \\ -\dfrac{\pi}{2}, & x = 0 \\ F(x + 2\pi). \end{cases}$

$\left(Ans. \quad -\dfrac{\pi}{4} + \sum_{n=1}^{\infty} \left(-\dfrac{1 - \cos n\pi}{\pi n^2} \cos nx + \dfrac{1 - 2\cos n\pi}{n} \sin nx \right) \right)$

(h) $F(x) = \begin{cases} \pi^2 - x^2, & -\pi \le x \le \pi \\ F(x + 2\pi). \end{cases}$

(i) $F(x) = \begin{cases} e^x, & -\pi < x < \pi \\ \cosh \pi, & x = \pi, -\pi \\ F(x + 2\pi). \end{cases}$

$\left(Ans. \quad \dfrac{2 \sinh \pi}{\pi} \left\{ \dfrac{1}{2} + \sum_{n=1}^{\infty} \left[\dfrac{(-1)^n}{1 + n^2} \cos nx + \dfrac{(-1)^{n+1} n}{1 + n^2} \sin nx \right] \right\} \right)$

(j) $F(x) = \begin{cases} \sin x + 3 \sin 2x + 2 \sin 3x, & -\pi \le x \le \pi \\ F(x + 2\pi). \end{cases}$

(k) $F(x) = \begin{cases} \cos x, & 0 < x < \pi \\ 0, & -\pi < x < 0 \\ \frac{1}{2}, & x = 0 \\ -\frac{1}{2}, & x = \pi, -\pi \\ F(x + 2\pi). \end{cases}$ $\left(Ans. \quad \frac{1}{2} \cos x + \dfrac{4}{\pi} \sum_{n=1}^{\infty} \dfrac{n}{4n^2 - 1} \sin 2nx \right)$

(l) $F(x) = \begin{cases} x^3, & -\pi < x < \pi \\ 0, & x = \pi, -\pi \\ F(x + 2\pi). \end{cases}$

(m) $F(x) = \begin{cases} 1, & 0 < x < \dfrac{\pi}{2} \\ 0, & \dfrac{\pi}{2} < x < \pi, -\pi < x < 0 \\ \frac{1}{2}, & x = -\pi, 0, \dfrac{\pi}{2}, \pi \\ F(x + 2\pi). \end{cases}$

$\left(Ans. \quad \dfrac{1}{4} + \dfrac{1}{\pi} \sum_{n=1}^{\infty} \left[\dfrac{\sin (n\pi/2)}{n} \cos nx + \dfrac{1 - \cos (n\pi/2)}{n} \sin nx \right] \right)$

2.3. Determine the even and odd periodic extensions and the Fourier sine and Fourier cosine series for each of the following functions:

(a) $f(x) = \sin x, \quad 0 < x < \pi$
(b) $f(x) = \cos x, \quad 0 < x < \pi$

(c) $f(x) = x,$ $\qquad 0 < x < \pi$
(d) $f(x) = x^2,$ $\qquad 0 < x < \pi$
(e) $f(x) = e^x,$ $\qquad 0 < x < \pi$

(f) $f(x) = \begin{cases} 2, & 0 < x < \dfrac{\pi}{2} \\[2mm] 3, & x = \dfrac{\pi}{2} \\[2mm] 4, & \dfrac{\pi}{2} < x < \pi. \end{cases}$

2.4. Each of the following functions is periodic and has the indicated period. Graph each function and find the convergent Fourier series representation indicated by Theorem 2.4.

(a) $F(x) = \begin{cases} -1, & -4 < x < 0 \\ 1, & 0 < x < 4 \\ 0, & x = 0, 4, -4 \\ F(x + 8). \end{cases}$ $\left(Ans. \quad \dfrac{4}{\pi} \sum_{n=1}^{\infty} \dfrac{1}{2n - 1} \sin \dfrac{(2n - 1)\pi x}{4} \right)$

(b) $F(x) = \begin{cases} 1, & -2 < x < 2 \\ -1, & 2 < x < 4, -4 < x < -2 \\ 0, & x = -4, -2, 2, 4 \\ F(x + 8). \end{cases}$ $\left(nAs. \quad \dfrac{4}{\pi} \sum_{n=1}^{\infty} \dfrac{(-1)^{n+1}}{2n - 1} \cos \dfrac{(2n - 1)\pi x}{4} \right)$

(c) $F(x) = \begin{cases} 0, & -2 < x < 0 \\ k, & 0 < x < 2 \\ \dfrac{k}{2}, & x = -2, 0, 2 \\ F(x + 4). \end{cases}$ $\left(Ans. \quad \dfrac{k}{2} + \dfrac{2k}{\pi} \sum_{n=1}^{\infty} \dfrac{1}{2n - 1} \sin \dfrac{(2n - 1)\pi x}{2} \right)$

(d) $F(x) = \begin{cases} |x|, & -L \le x \le L, L > 0 \\ F(x + 2L). \end{cases}$
$\left(Ans. \quad \dfrac{L}{2} - \dfrac{4L}{\pi^2} \sum_{n=1}^{\infty} \dfrac{1}{(2n - 1)^2} \cos \dfrac{(2n - 1)\pi x}{L} \right)$

2.5. Show that the following series is not a Fourier series:

$$\sum_{n=1}^{\infty} \frac{\sin nx}{\log (1 + n)}.$$

2.6. By means of Fourier series, prove each of the following:

(a) $1 + \dfrac{1}{2^2} + \dfrac{1}{3^2} + \cdots + \dfrac{1}{n^2} + \cdots = \dfrac{\pi^2}{6}$

(b) $1 + \dfrac{1}{3^2} + \dfrac{1}{5^2} + \cdots + \dfrac{1}{(2n - 1)^2} + \cdots = \dfrac{\pi^2}{8}$

(c) $1 + \dfrac{1}{2^4} + \dfrac{1}{3^4} + \dfrac{1}{4^4} + \cdots + \dfrac{1}{n^4} + \cdots = \dfrac{\pi^4}{90}$

(d) $1 - \dfrac{1}{2^2} + \dfrac{1}{3^2} - \dfrac{1}{4^2} + \cdots + \dfrac{(-1)^{n+1}}{n^2} + \cdots = \dfrac{\pi^2}{12}.$

2.7. Determine whether or not each series of Exercise 2.2 satisfies the conditions of Theorem 2.5.   In each case where the conditions are satisfied, carry out the differentiation.

2.8. Determine whether or not each series of Exercises 3 and 4 is termwise differentiable.   In each case possible carry out the differentiation.

2.9. Show that if $F(x)$ is sectionally continuous on $-\pi \le x \le \pi$, $F(x + 2\pi) = F(x)$, and series (2.24) is the Fourier series representation of $F(x)$ for all real x, then termwise integration is a valid procedure, that is, for fixed α and all x in $-\pi \le x \le \pi$,

$$\int_\alpha^x F(t)\, dt \equiv \int_\alpha^x \frac{a_0}{2}\, dt + \sum_{n=1}^\infty \left(\int_\alpha^x a_n \cos nt\, dt + \int_\alpha^x b_n \sin nt\, dt \right).$$

2.10. Let a, b be two real numbers with $a < b$.   Let M be the set of all real numbers x which satisfy $a \le x \le b$.   If $F(x)$ is sectionally continuous on M, then show that $\lim\limits_{n \to \infty} \displaystyle\int_a^b F(x) \cos nx\, dx = 0$.

2.11. Let $F(x)$ be defined for all real x, have period 2π, and be piecewise continuous on $-\pi \le x \le \pi$.   Let $S_n(x)$ be given by

$$S_n(x) = \frac{\alpha_0}{2} + \sum_{k=1}^n (\alpha_n \cos kx + \beta_n \sin kx).$$

$S_n(x)$ is said to be a best approximation in the mean to $F(x)$ provided

$$\frac{1}{2\pi} \int_{-\pi}^\pi [F(x) - S_n(x)]^2\, dx$$

is a minimum.   Show that in general $S_n(x)$ is a best approximation in the mean to $F(x)$ if α_n, β_n are given by

$$\alpha_n = a_n = \frac{1}{\pi} \int_{-\pi}^\pi F(x) \cos nx\, dx, \qquad \beta_n = b_n = \frac{1}{\pi} \int_{-\pi}^\pi F(x) \sin nx\, dx.$$

2.12. Let $F(x)$ be defined for all real x by

$$F(x) = \begin{cases} 7 + \dfrac{5(x + 1)}{\pi - 1}, & -\pi \le x \le -1 \\[2mm] 7 - \dfrac{8(x + 1)}{3}, & -1 \le x \le 2 \\[2mm] -1 + \dfrac{3(x - 2)}{\pi - 2}, & 2 \le x \le \pi \\[2mm] F(x + 2\pi). \end{cases}$$

Find a finite trigonometric sum $S_N(x)$, of the form described in Theorem 2.7, such that for all x

$$|F(x) - S_N(x)| < 0.1.$$

2.13. (*The Weierstrass Approximation Theorem*) Show that any given function $F(x)$ which is *continuous* on $-\pi \le x \le \pi$ can be approximated uniformly on $-\pi \le x \le \pi$ by a finite polynomial sum of the type

$$P_n(x) = \sum_{j=0}^{n} c_j x^j$$

with any preassigned degree of accuracy; that is, given any $\epsilon > 0$, there exists $P_n(x)$ such that, for all x in $-\pi \le x \le \pi$,

$$|P_n(x) - F(x)| < \epsilon.$$

2.14. Let B be the square with vertices $(0,0)$, $(\pi,0)$, (π,π), $(0,\pi)$ and let R be its interior. Set $\bar{R} = R \cup B$. Then give three examples of series of the form

$$\sum_{n=1}^{\infty} F_n(x)G_n(y)$$

which, by Abel's test, converge uniformly on $\bar{R}$ in the two variables x and y together.

2.15. Generalize Definition 2.1 so that a function may be periodic even though it is not defined for all real x, e.g., $\tan x$. Show how to develop a theory of Fourier series for such functions.

2.16. For $a < b$, let M be the set of real numbers x such that $a \le x \le b$. Show that if $F(x)$ has a derivative at each point of M and if $F'(x)$ is piecewise continuous on M, then $F'(x)$ is continuous on M.

Chapter 3

SECOND-ORDER PARTIAL DIFFERENTIAL EQUATIONS

3.1 Partial Differential Equations

Definition 3.1. A partial differential equation in the dependent variable u and the independent variables x, y is an equation which is, or can be put, in the form

$$\mathscr{F}(x,y,u,u_x,u_y,u_{xx},u_{xy},u_{yy},u_{xxx},u_{xxy}, \ldots) = 0, \tag{3.1}$$

where $\mathscr{F}$ is a function of the indicated quantities and at least one partial derivative occurs.

Example 1. $u_x = u + x + y$ is a partial differential equation.

Example 2. $u_{xx} - u_{yy} = 0$ is a partial differential equation.

Example 3. $\dfrac{\partial^3 u}{\partial x^3} + 3y\dfrac{\partial^2 u}{\partial x^2} - x\dfrac{\partial u}{\partial x}\dfrac{\partial^2 u}{\partial y^2} + u\left(\dfrac{\partial u}{\partial y}\right)^3 = e^{xyu}$ is a partial differential equation.

The usual abbreviation for *partial differential equation* is p.d.e.

Probably the class of partial differential equations which has been of most interest is the class of second-order equations.

Definition 3.2. A p.d.e. in dependent variable u and independent variables x, y is said to be a *second-order quasi-linear* p.d.e. on a given plane point set M if and only if it is, or can be put, in the form

$$Au_{xx} + Bu_{xy} + Cu_{yy} + \Phi(x,y,u,u_x,u_y) = 0, \tag{3.2}$$

where A, B, C are functions of x and y on M, $A^2 + B^2 + C^2$ is never zero on M, and Φ is a function of the indicated quantities.

Example 1. $u_{xx} + xu_{yy} - u_x - e^{xy}\sin u = 0$ is a second-order quasi-linear p.d.e. on E^2.

Example 2. Let $(x,y) \in M$ if and only if $x^2 + y^2 < 1$. Then $u_{xx} - \sqrt{1 - x^2 - y^2} \, u_{yy} = 0$ is a second-order quasi-linear p.d.e. on M.

Definition 3.3. A second-order quasi-linear p.d.e. in dependent variable u and independent variables x, y is said to be linear on a given plane point set M if and only if it is, or can be put, in the form

$$Au_{xx} + Bu_{xy} + Cu_{yy} + Du_x + Eu_y + Fu + G = 0, \qquad (3.3)$$

where A, B, C, D, E, F, G are functions of x and y on M and $A^2 + B^2 + C^2$ is never zero on M.

Example 1. On E^2, $u_{xx} - u_{yy} + u = 0$ is a linear, second-order p.d.e.

Example 2. On E^2, $\dfrac{\partial^2 u}{\partial x^2} - \dfrac{\partial^2 u}{\partial y^2} + x^2 \dfrac{\partial u}{\partial x} - xy \dfrac{\partial u}{\partial y} + u = 1$ is a linear, second-order p.d.e.

Example 3. Let $(x,y) \in M$ if and only if $x^2 + y^2 < 1$. Then $u_{xx} + \sqrt{1 - x^2 - y^2} \, u_{xy} - y^3 u_x - xyu + e^{xy} = 0$ is a linear, second-order p.d.e. on M.

Of course, since (3.3) is a special case of (3.2), any result established for (3.2) is also valid for (3.3).

Definition 3.4. A function $u = f(x,y)$ is said to be a solution of (3.2) on plane point set M if and only if, on M, $f(x,y) \in C^2$ and the function and its derivatives satisfy the equation identically in x and y.

Example 1. Let $(x,y) \in M$ if and only if $x^2 + y^2 < 1$. Then

$$u_{xy} + u_x = -\frac{xy + xu^2}{(1 - x^2 - y^2)^{3/2}} \qquad (3.4)$$

is a quasi-linear p.d.e. on M. The function

$$f(x,y) = \sqrt{1 - x^2 - y^2} \qquad (3.5)$$

is a solution of (3.4) on M since $f(x,y) \in C^2$ on M and since (3.5), together with $f_x = -\dfrac{x}{(1 - x^2 - y^2)^{1/2}}$, $f_{xy} = -\dfrac{xy}{(1 - x^2 - y^2)^{3/2}}$, yield upon substitution into (3.4) the identity

$$\frac{-xy}{(1 - x^2 - y^2)^{3/2}} + \frac{-x}{(1 - x^2 - y^2)^{1/2}} \equiv -\frac{xy + x(1 - x^2 - y^2)}{(1 - x^2 - y^2)^{3/2}} \,.$$

Example 2. On E^2,

$$\frac{\partial^2 u}{\partial x^2} + \frac{\partial^2 u}{\partial y^2} = 0 \qquad (3.6)$$

is a linear, second-order p.d.e. The function

$$f(x,y) = x^2 - y^2 \qquad (3.7)$$

is a solution of (3.6) on E^2 since $f(x,y) \in C^2$ on E^2 and since (3.7) implies $\partial^2 u/\partial x^2 = 1$, $\partial^2 u/\partial y^2 = -1$, which yield upon substitution into (3.6) the identity

$$1 - 1 \equiv 0.$$

Note that one of the ramifications of the assumption $f(x,y) \in C^2$ in Definition 3.4 is that $u_{xy} \equiv u_{yx}$, so that one need not be concerned about the order of differentiation.

Consider now a particular representation of p.d.e. (3.2). On E^2, consider, say,

$$u_{xx} - u_{yy} = 0. \qquad (3.8)$$

Various solutions of (3.8) on E^2 are

$$f(x,y) = (x + y)^3$$

$$f(x,y) = \sin(x - y)$$

$$f(x,y) = (x + y)^{27} + \cos(x - y)\sin(x - y) + e^{3(x+y)}$$

$$f(x,y) = (x + y)^2 \sin(x + y) + (x - y)^2 e^{x-y}.$$

As a matter of fact, direct calculation reveals that

$$f(x,y) = f_1(x + y) + f_2(x - y), \qquad (3.9)$$

where f_1, f_2 are arbitrary functions of class C^2, is a solution of (3.8) on E^2. From the point of view that p.d.e. (3.8) is a *second*-order equation, the fact that solution (3.9) contains *two* arbitrary functions is somewhat satisfying. However, the fact that f_1, f_2 are quite arbitrary indicates that solution (3.9) may be far too general to be of any practical value. Similar solutions in terms of two arbitrary functions can readily be produced for other second-order quasi-linear partial differential equations. We shall therefore seek next to impose additional conditions on the solution of a differential equation in order to insure the uniqueness of that solution.

3.2 The Initial-value Problem

Attention will now be directed to Eqs. (3.2) and (3.3). In order to cultivate some intuition about what is reasonable and what is unreasonable in any discussion of quasi-linear, and, as a special case, linear, partial differential equations of second order, it is advantageous to examine first some aspects of the theory of second-order *ordinary* differential equations.

For the second-order ordinary differential equation

$$\frac{d^2y}{dx^2} + \phi\left(x, y, \frac{dy}{dx}\right) = 0, \qquad (3.10)$$

a problem of major concern is the *initial-value* problem, that is, given a fixed value $\bar{x}$ and two fixed constants a_1, a_2, one must find on a given set of real numbers a solution $y = g(x)$ of (3.10) such that $g(\bar{x}) = a_1$, $g'(\bar{x}) = a_2$. Geometrically, an initial-value problem requires that one find on a given set of real numbers a solution $y = g(x)$ of (3.10) such that the point $(\bar{x}, a_1)$ lies on the graph of $y = g(x)$ while the tangent line to $y = g(x)$ at $(\bar{x}, a_1)$ has slope a_2.

In order to consider an analogous problem for (3.2), one must observe that an obvious difference between a solution $u = f(x,y)$ of (3.2) and a solution $y = g(x)$ of (3.10) is that $f(x,y)$ is a function of two independent variables while $g(x)$ is a function of only one independent variable. This suggests that the roles played by *point*, *plane*, and *tangent line* for (3.10) be replaced when considering (3.2) by corresponding geometric entities of one higher dimension, that is, by *curve*, *three space*, and *tangent plane*, respectively. One might surmise then, in a geometric fashion, that an initial-value problem for p.d.e. (3.2) is a problem in which one is asked to find a solution

$$u = f(x,y) \qquad (3.11)$$

on a given plane point set M, of (3.2), such that:

    (a') a fixed space curve C^* lies on the graph of (3.11), and
    (b') at each point of C^*, the graph of (3.11) has a prescribed tangent plane.

This formulation of an initial-value problem will now be considered in greater detail.

An alternate way, which is quite often convenient, of considering condition (a'), above, is to consider the projection C_0 of C^* in the xy plane and to prescribe $f(x,y)$ on C_0. Thus, an initial-value problem for p.d.e. (3.2) requires one to find a solution (3.11) of (3.2) on a given plane point set M such that:

    (a") $f(x,y)$ assumes prescribed values at each point of the xy projection C_0 of fixed space curve C^*, and
    (b") at each point of C^*, the graph of (3.11) has a prescribed tangent plane.

We next seek to rephrase analytically and to refine conditions (a'') and (b''). With respect to (a''), let N be an interval of real numbers (where the real line is classified as such an interval) and let C^* be a space curve with parametric representation

$$C^*: x = x(t), \, y = y(t), \, u = u(t); \qquad t \in N. \qquad (3.12)$$

Without loss of generality, assume $t \in N$ implies $(x(t), y(t)) \in M$. In these elementary considerations it is desirable that the curve C^* be "well behaved." This is accomplished by assuming $x(t)$, $y(t)$, $u(t) \in C^2$ and that $\left(\dfrac{dx}{dt}\right)^2 + \left(\dfrac{dy}{dt}\right)^2 + \left(\dfrac{du}{dt}\right)^2 \neq 0$ for any $t \in N$. Then (3.12) may be replaced by

$$\left. \begin{array}{l} C^*: x = x(t), \, y = y(t), \, u = u(t); \qquad x(t), \, y(t), \, u(t) \in C^2; \\[2mm] \left(\dfrac{dx}{dt}\right)^2 + \left(\dfrac{dy}{dt}\right)^2 + \left(\dfrac{du}{dt}\right)^2 \neq 0; \qquad t \in N. \end{array} \right\} \qquad (3.13)$$

Concerning the projection C_0 of C^*, it is therefore desirable that C_0 should have the parametric representation

$$\left. \begin{array}{l} C_0: x = x(t), \, y = y(t), \, u = 0; \qquad x(t), \, y(t) \in C^2; \\[2mm] \left(\dfrac{dx}{dt}\right)^2 + \left(\dfrac{dy}{dt}\right)^2 \neq 0, \qquad t \in N. \end{array} \right\} \qquad (3.14)$$

With respect to (b''), note that prescribing the tangent plane at various points of (3.11) is equivalent to prescribing f_x and f_y at these points.

The above heuristic discussion culminates finally in the following definition of an initial-value problem.

Definition 3.5. An initial-value problem for quasi-linear p.d.e. (3.2) is a problem in which one is asked to find a solution

$$u = f(x, y) \qquad (3.11)$$

of (3.2) on a given plane point set M such that if N is an interval of real numbers, C^* is a space curve given by (3.13), C_0 is the xy projection of C^* and is given by (3.14), and $t \in N$ implies $(x(t), y(t)) \in M$, then

(a) f assumes prescribed values at each point of C_0, and
(b) f_x, f_y assume prescribed values at each point of C_0.

The following is an example of an initial-value problem.

Example. If N is the set of all real numbers, let C^* be defined by $C^*: x = t,\, y = t^2,\, u = t^3,\, t \in N$. At each point of C_0, prescribe f, f_x, f_y by $f = t^3,\, f_x = 2t - 1,\, f_y = 1 - t^2$. Then, if M is E^2, find a solution $u = f(x,y)$ on M of

$$\frac{\partial^2 u}{\partial x^2} - \frac{\partial^2 u}{\partial y^2} = 0$$

such that

(a) $t^3 \equiv f(t,t^2),\qquad t \in N,$

(b) $\dfrac{\partial f(t,t^2)}{\partial x} \equiv 2t - 1,\qquad \dfrac{\partial f(t,t^2)}{\partial y} \equiv 1 - t^2,\qquad t \in N.$

With some further thought, it becomes apparent that an initial-value problem for partial differential equations implies a compatibility condition which is not implied for ordinary differential equations. With regard to the graph of (3.11), $[f_x, f_y, -1]$ are direction numbers of a normal vector and hence at any point determine the tangent plane. However, since (3.13) is to be on the graph of (3.11), it is necessary that at any point of C^* the tangent vector to C^* lie in the tangent plane at that point of the graph of (3.11). Thus, if $u = f(x,y)$ is a solution of the initial-value problem described in Definition 3.5, it is necessary that

$$\frac{\partial f(x(t),y(t))}{\partial x}\frac{dx(t)}{dt} + \frac{\partial f(x(t),y(t))}{\partial y}\frac{dy(t)}{dt} + (-1)\frac{du(t)}{dt} \equiv 0, \qquad t \in N,$$

or, equivalently, that

$$\frac{du(t)}{dt} \equiv \frac{\partial f(x(t),y(t))}{\partial x}\frac{dx(t)}{dt} + \frac{\partial f(x(t),y(t))}{\partial y}\frac{dy(t)}{dt}, \qquad t \in N.$$

This latter identity is called the *strip condition* and is analytically nothing more than the chain rule.

3.3   Characteristics

Further heuristic considerations of initial-value problems lead quite naturally to the very basic concept of a *characteristic*, or a *characteristic curve*. For this purpose, assume that a given initial-value problem has a *unique* solution $u = f(x,y)$ on plane point set M; that for some parameter value $t = \bar{t}$, the point $(\bar{x},\bar{y})$, where $\bar{x} = x(\bar{t})$, $\bar{y} = y(\bar{t})$, is not only a point of C_0 but is also an interior point of M; and that $f(x,y)$ has a Taylor series expansion which is valid in some circular neighborhood of $(\bar{x},\bar{y})$. Then, on this circular neighborhood,

$$f(x,y) \equiv f(\bar{x},\bar{y}) + \left[(x - \bar{x})\frac{\partial f(\bar{x},\bar{y})}{\partial x} + (y - \bar{y})\frac{\partial f(\bar{x},\bar{y})}{\partial y} \right]$$

$$+ \frac{1}{2!}\left[(x - \bar{x})^2 \frac{\partial^2 f(\bar{x},\bar{y})}{\partial x^2} + 2(x - \bar{x})(y - \bar{y})\frac{\partial^2 f(\bar{x},\bar{y})}{\partial x\, \partial y} + (y - \bar{y})^2 \frac{\partial^2 f(\bar{x},\bar{y})}{\partial y^2} \right]$$

$$+ \cdots$$

$$+ \frac{1}{n!}\left[(x - \bar{x})^n \frac{\partial^n f(\bar{x},\bar{y})}{\partial x^n} + C_{n,1}(x - \bar{x})^{n-1}(y - \bar{y})\frac{\partial^n f(\bar{x},\bar{y})}{\partial x^{n-1}\, \partial y} + \cdots \right.$$

$$\left. + C_{n,n}(y - \bar{y})^n \frac{\partial^n f(\bar{x},\bar{y})}{\partial y^n} \right] + \cdots. \quad (3.15)$$

If one could calculate all the Taylor coefficients in (3.15), then the given initial-value problem would be solved. From the data given in an initial-value problem, it follows that $f(\bar{x},\bar{y})$, $\dfrac{\partial f(\bar{x},\bar{y})}{\partial x}$, $\dfrac{\partial f(\bar{x},\bar{y})}{\partial y}$ are easily determined, so that the first few terms of (3.15) are known. One might then attempt to calculate the next three Taylor coefficients by determining $\dfrac{\partial^2 f(\bar{x},\bar{y})}{\partial x^2}$, $\dfrac{\partial^2 f(\bar{x},\bar{y})}{\partial x\, \partial y}$, $\dfrac{\partial^2 f(\bar{x},\bar{y})}{\partial y^2}$ in terms of quantities which are already known as follows. Since $f(x,y)$ is a solution of (3.2) on M, one has

$$A \frac{\partial^2 f(\bar{x},\bar{y})}{\partial x^2} + B \frac{\partial^2 f(\bar{x},\bar{y})}{\partial x\, \partial y} + C \frac{\partial^2 f(\bar{x},\bar{y})}{\partial y^2} = -\Phi\left(\bar{x},\bar{y},f(\bar{x},\bar{y}), \frac{\partial f(\bar{x},\bar{y})}{\partial x}, \frac{\partial f(\bar{x},\bar{y})}{\partial y} \right).$$
$$(3.16)$$

Moreover, since

$$\frac{d}{dt}\left(\frac{\partial f}{\partial x}\right) \equiv \frac{\partial}{\partial x}\left(\frac{\partial f}{\partial x}\right)\frac{dx}{dt} + \frac{\partial}{\partial y}\left(\frac{\partial f}{\partial x}\right)\frac{dy}{dt},$$

$$\frac{d}{dt}\left(\frac{\partial f}{\partial y}\right) \equiv \frac{\partial}{\partial x}\left(\frac{\partial f}{\partial y}\right)\frac{dx}{dt} + \frac{\partial}{\partial y}\left(\frac{\partial f}{\partial y}\right)\frac{dy}{dt},$$

one must have

$$\frac{dx(\bar{t})}{dt}\frac{\partial^2 f(\bar{x},\bar{y})}{\partial x^2} + \frac{dy(\bar{t})}{dt}\frac{\partial^2 f(\bar{x},\bar{y})}{\partial x\, \partial y} = \left[\frac{d}{dt}\left(\frac{\partial f}{\partial x}\right)\right]\bigg|_{t=\bar{t}}, \quad (3.17)$$

$$\frac{dx(\bar{t})}{dt}\frac{\partial^2 f(\bar{x},\bar{y})}{\partial x\, \partial y} + \frac{dy(\bar{t})}{dt}\frac{\partial^2 f(\bar{x},\bar{y})}{\partial y^2} = \left[\frac{d}{dt}\left(\frac{\partial f}{\partial y}\right)\right]\bigg|_{t=\bar{t}}. \quad (3.18)$$

However, all quantities of (3.16) to (3.18) are known except $\dfrac{\partial^2 f(\bar{x},\bar{y})}{\partial x^2}$, $\dfrac{\partial^2 f(\bar{x},\bar{y})}{\partial x \, \partial y}$, $\dfrac{\partial^2 f(\bar{x},\bar{y})}{\partial y^2}$. Hence, (3.16) to (3.18) constitute a linear algebraic system of equations in the three desired second-order derivatives.

Since we are seeking a *unique* solution of an initial-value problem, the Taylor coefficients must be unique. With respect to the second-order derivatives under consideration, it is desirable that the determinant of the linear algebraic system (3.16) to (3.18) be nonzero, that is,

$$\begin{vmatrix} A & B & C \\ \dfrac{dx}{dt} & \dfrac{dy}{dt} & 0 \\ 0 & \dfrac{dx}{dt} & \dfrac{dy}{dt} \end{vmatrix} \neq 0.$$

Hence, in seeking a unique solution of type (3.15), one would wish to *avoid* choosing C_0 to be a curve which allows

$$\begin{vmatrix} A & B & C \\ \dfrac{dx}{dt} & \dfrac{dy}{dt} & 0 \\ 0 & \dfrac{dx}{dt} & \dfrac{dy}{dt} \end{vmatrix} = 0,$$

or, equivalently, which allows

$$A(dy)^2 - B(dy)(dx) + C(dx)^2 = 0.$$

Definition 3.6. The *characteristic differential equation* associated with quasi-linear second-order p.d.e. (3.2) is

$$A(dy)^2 - B(dy)(dx) + C(dx)^2 = 0. \tag{3.19}$$

Definition 3.7. The *characteristic curves*, or, *characteristics*, associated with quasi-linear, second-order p.d.e. (3.2) are the solutions of the associated characteristic differential equation.

Example. On E^2, the characteristic curves associated with

$$u_{xx} - u_{yy} = 0$$

are defined by the characteristic differential equation

$$(dy)^2 - (dx)^2 = 0,$$

or, equivalently, by

$$\left(\frac{dy}{dx}\right)^2 - 1 = 0,$$

and are therefore given by

$$y = x + c_1, \qquad y = -x + c_2,$$

where c_1, c_2 are arbitrary constants.

3.4 Canonical Forms

One can fruitfully explore, with the aid of the concept of a characteristic, the question of making a change of variables so as to simplify p.d.e. (3.3). It will be sufficient for our purposes to confine ourselves in the remainder of the chapter to the linear p.d.e.

$$Au_{xx} + Bu_{xy} + Cu_{yy} + Du_x + Eu_y + Fu + G = 0, \qquad (3.20)$$

where A, B, C, D, E, F are real *constants*, $A^2 + B^2 + C^2 \neq 0$, and $G = G(x,y)$ is a real-valued function defined on a given plane point set M.

Associated with (3.20) is the characteristic differential equation

$$A(dy)^2 - B(dy)(dx) + C(dx)^2 = 0, \qquad (3.21)$$

where A, B, C are constants and $A^2 + B^2 + C^2 \neq 0$. For intuitive purposes, suppose none of A, B, C is zero. The case where at least one of A, B, C is zero will follow readily in the more precise considerations to be given later. If none of A, B, C is zero, then, set

$$\frac{dy}{dx} = -\lambda, \qquad (3.22)$$

so that (3.21) is equivalent to

$$A\lambda^2 + B\lambda + C = 0. \qquad (3.23)$$

Since (3.23) is quadratic in λ and has real, nonzero coefficients, it follows that (3.23) has two roots λ_1, λ_2 given by

$$\lambda_1 = \frac{-B + \sqrt{B^2 - 4AC}}{2A}, \qquad \lambda_2 = \frac{-B - \sqrt{B^2 - 4AC}}{2A}. \qquad (3.24)$$

Since the pair of roots λ_1, λ_2 must fall into one of the three categories (a) λ_1, λ_2 real and unequal, (b) λ_1, λ_2 conjugate complex, (c) λ_1, λ_2 real and equal, it follows, correspondingly, that (3.20) always belongs to one of three categories according as (a) $B^2 - 4AC > 0$, (b) $B^2 - 4AC < 0$, (c) $B^2 - 4AC = 0$. By analogy with the well-known classifications of analytical geometry that

$$Ax^2 + Bxy + Cy^2 + Dx + Ey + F = 0, \qquad (A^2 + B^2 + C^2 \neq 0)$$

is   (3.25)

(a) of hyperbolic type if and only if $B^2 - 4AC > 0$,
(b) of elliptic type if and only if $B^2 - 4AC < 0$,
(c) of parabolic type if and only if $B^2 - 4AC = 0$,

one gives the following definition.

Definition 3.8.   Given the second-order linear p.d.e.

$$Au_{xx} + Bu_{xy} + Cu_{yy} + Du_x + Eu_y + Fu + G = 0, \qquad (3.26)$$

where A, B, C, D, E, F are real constants, $A^2 + B^2 + C^2 \neq 0$, and $G = G(x,y)$ is a real-valued function defined on a given plane point set M, then on M p.d.e. (3.26) is said to be:

(a) hyperbolic if and only if $B^2 - 4AC > 0$,
(b) elliptic if and only if $B^2 - 4AC < 0$,
(c) parabolic if and only if $B^2 - 4AC = 0$.

Example 1.   On E^2, $u_{xx} - u_{yy} = 0$ is hyperbolic since $B^2 - 4AC = 4$.

Example 2.   On E^2, $u_{xx} + u_{yy} - u_x + u = 0$ is elliptic since $B^2 - 4AC = -4$.

Example 3.   On E^2, $\dfrac{\partial^2 u}{\partial x^2} - \dfrac{\partial u}{\partial y} + \dfrac{\partial u}{\partial x} + e^{xy} = 0$ is parabolic since $B^2 - 4AC = 0$.

Two basic results of analytical geometry germane to (3.25) are now worthy of recall.   First, the quantity $B^2 - 4AC$ is invariant under translation and rotation of axes.   Second, there always exists a rotation so that in the new coordinate system (3.25) simplifies to a form in which at least one of the three second-degree terms is not present.   We now investigate in what fashion these results can be extended to p.d.e. (3.26).

Definition 3.9.   The algebraic transformation, or mapping,

$$\left. \begin{aligned} \bar{x} &= \alpha_1 x + \beta_1 y + \gamma_1 \\ \bar{y} &= \alpha_2 x + \beta_2 y + \gamma_2, \end{aligned} \right\} \qquad (3.27)$$

where $\alpha_1, \beta_1, \gamma_1, \alpha_2, \beta_2, \gamma_2$ are constants and $\alpha_1\beta_2 - \alpha_2\beta_1 \neq 0$ is called an affine transformation or affine mapping.

Example 1.   Rotations and translations in E^2 are affine transformations.

Example 2. $\begin{cases} \bar{x} = x + y \\ \bar{y} = x - y + 2 \end{cases}$
defines an affine mapping.

Note that the condition $\alpha_1\beta_2 - \alpha_2\beta_1 \neq 0$ merely prescribes that the Jacobian of the transformation be nonzero.

Definition 3.10.   The quantity $B^2 - 4AC$ is called the *discriminant* of linear, second-order p.d.e. (3.26).

Theorem 3.1.   The *sign* of the discriminant $B^2 - 4AC$ of linear, second-order p.d.e. (3.26) is invariant under the general affine transformation (3.27).

Proof.   Under mapping (3.27), p.d.e. (3.26) becomes

$$\bar{A}u_{\bar{x}\bar{x}} + \bar{B}u_{\bar{x}\bar{y}} + \bar{C}u_{\bar{y}\bar{y}} + \bar{D}u_{\bar{x}} + \bar{E}u_{\bar{y}} + \bar{F}u + \bar{G} = 0, \qquad (3.28)$$

where

$$\bar{A} = \alpha_1^2 A + \alpha_1\beta_1 B + \beta_1^2 C, \qquad (3.29)$$

$$\bar{B} = 2\alpha_1\alpha_2 A + \alpha_1\beta_2 B + \alpha_2\beta_1 B + 2\beta_1\beta_2 C, \qquad (3.30)$$

$$\bar{C} = \alpha_2^2 A + \alpha_2\beta_2 B + \beta_2^2 C, \qquad (3.31)$$

$$\bar{D} = \alpha_1 D + \beta_1 E, \qquad (3.32)$$

$$\bar{E} = \alpha_2 D + \beta_2 E, \qquad (3.33)$$

$$\bar{F} = F, \qquad (3.34)$$

$$\bar{G} = \bar{G}(\bar{x},\bar{y}) = G(x,y). \qquad (3.35)$$

Unexciting calculation reveals

$$\bar{B}^2 - 4\bar{A}\bar{C} = (2\alpha_1\alpha_2 A + \alpha_1\beta_2 B + \alpha_2\beta_1 B + 2\beta_1\beta_2 C)^2$$

$$- 4(\alpha_1^2 A + \alpha_1\beta_1 B + \beta_1^2 C)(\alpha_2^2 A + \alpha_2\beta_2 B + \beta_2^2 C)$$

$$= (B^2 - 4AC)(\alpha_1\beta_2 - \alpha_2\beta_1)^2.$$

Since $\alpha_1\beta_2 - \alpha_2\beta_1 \neq 0$, the theorem follows from this latter result.

Corollary.   Under an affine transformation, a hyperbolic, elliptic, or parabolic equation transforms, respectively, into a hyperbolic, elliptic, or parabolic equation.

The question of simplifying p.d.e. (3.26) by an affine transformation so that at least one of $\bar{A}, \bar{B}, \bar{C}$ in (3.28) is zero can be resolved with the aid of the theory of quadratic forms (Tamarkin and Feller [47]).   We shall, however, use an alternate approach which explores the possibility of making a transformation so that the characteristic curves of the given partial differential equation map into the coordinate curves of the new coordinate system.

Theorem 3.2.   Given linear p.d.e.

$$Au_{xx} + Bu_{xy} + Cu_{yy} + Du_x + Eu_y + Fu + G = 0, \qquad (3.36)$$

where A, B, C, D, E, F are real constants, G is a real-valued function of x and y on plane point set M, and $A^2 + B^2 + C^2 \neq 0$, then if (3.36) is hyperbolic on M there exists an affine transformation of the form

$$\left.\begin{array}{l} \bar{x} = \alpha_1 x + \beta_1 y \\ \bar{y} = \alpha_2 x + \beta_2 y \end{array}\right\} \tag{3.37}$$

such that in the $\bar{x}\bar{y}$ coordinates, (3.36) has the form

$$u_{\bar{x}\bar{y}} = D_1 u_{\bar{x}} + E_1 u_{\bar{y}} + F_1 u + G_1(\bar{x},\bar{y}), \tag{3.38}$$

where D_1, E_1, F_1 are constants and G_1 is a real-valued function of $\bar{x}$, $\bar{y}$ on the set M_1 which is the map of M under (3.37).

Proof. Consider the problem of determining α_1, β_1, α_2, β_2 for (3.37) so that in the $\bar{x}\bar{y}$ coordinates the equations of the characteristic curves of (3.36) are $\bar{x} = c_1$, $\bar{y} = c_2$, where c_1, c_2 are constants. This $\bar{x}\bar{y}$ system, if it exists, will be called a characteristic coordinate system.

Suppose first that *none* of A, B, C is zero. The characteristics of (3.36) can be found from

$$\frac{dy}{dx} = -\lambda_1, \qquad \frac{dy}{dx} = -\lambda_2, \tag{3.39}$$

where λ_1, λ_2 are given by (3.24). Since $B^2 - 4AC > 0$, λ_1, λ_2 are both real and distinct, so that (3.24) and (3.39) imply that the equations of the characteristics are

$$\left.\begin{array}{l} y + \dfrac{-B + \sqrt{B^2 - 4AC}}{2A}\, x = c_1 \\[3mm] y + \dfrac{-B - \sqrt{B^2 - 4AC}}{2A}\, x = c_2, \end{array}\right\} \tag{3.40}$$

where c_1, c_2 are arbitrary constants. Since $\bar{x} = c_1$, $\bar{y} = c_2$ are equations of coordinate lines in the $\bar{x}\bar{y}$ coordinate system, the affine transformation

$$\left.\begin{array}{l} \bar{x} = \dfrac{-B + \sqrt{B^2 - 4AC}}{2A}\, x + y \\[3mm] \bar{y} = \dfrac{-B - \sqrt{B^2 - 4AC}}{2A}\, x + y \end{array}\right\} \tag{3.41}$$

is motivated, for under this transformation the characteristics with Eqs. (3.40) map into the coordinate lines of the $\bar{x}\bar{y}$ system.

Mapping (3.41) implies then with respect to (3.37) that

$$\alpha_1 = \frac{-B + \sqrt{B^2 - 4AC}}{2A}, \quad \alpha_2 = \frac{-B - \sqrt{B^2 - 4AC}}{2A}, \quad \beta_1 = \beta_2 = 1$$

and (3.29) to (3.35) imply

$$\bar{A} = \bar{C} = 0, \quad \bar{B} = \frac{B^2 - 4AC}{-A}, \quad \bar{D} = \frac{-B + \sqrt{B^2 - 4AC}}{2A} D + E,$$

$$\bar{E} = \frac{-B - \sqrt{B^2 - 4AC}}{2A} D + E, \quad \bar{F} = F, \quad \bar{G} = \bar{G}(\bar{x}, \bar{y}) = G(x, y).$$

Thus (3.36) transforms into

$$\frac{B^2 - 4AC}{-A} u_{\bar{x}\bar{y}} + \left(\frac{-B + \sqrt{B^2 - 4AC}}{2A} D + E \right) u_{\bar{x}}$$

$$+ \left(\frac{-B - \sqrt{B^2 - 4AC}}{2A} D + E \right) u_{\bar{y}} + \bar{F} u + \bar{G} = 0. \quad (3.42)$$

Since $A \neq 0$ and $B^2 - 4AC \neq 0$, (3.42) is equivalent to

$$u_{\bar{x}\bar{y}} = \frac{(-B + \sqrt{B^2 - 4AC})D + 2EA}{2(B^2 - 4AC)} u_{\bar{x}}$$

$$+ \frac{(-B - \sqrt{B^2 - 4AC})D + 2EA}{2(B^2 - 4AC)} u_{\bar{y}} + \frac{\bar{F}A}{B^2 - 4AC} u + \frac{\bar{G}A}{B^2 - 4AC}.$$

$$(3.42')$$

Setting

$$D_1 = \frac{(-B + \sqrt{B^2 - 4AC})D + 2EA}{2(B^2 - 4AC)}, \quad F_1 = \frac{\bar{F}A}{B^2 - 4AC}$$

$$E_1 = \frac{(-B - \sqrt{B^2 - 4AC})D + 2EA}{2(B^2 - 4AC)}, \quad G_1 = \frac{\bar{G}A}{B^2 - 4AC}$$

and substituting into (3.42') readily yields (3.38). Thus, if none of A, B, C is zero, the theorem is proved.

Suppose now *exactly one* of A, B, C is zero. If $B = 0$, or $C = 0$, the above considerations apply and the theorem is valid. If $A = 0$, $B \neq 0$, $C \neq 0$, then the characteristic differential equation is

$$-B \, dy \, dx + C(dx)^2 = 0. \quad (3.43)$$

Setting

$$\frac{dx}{dy} = -\lambda \quad (3.44)$$

implies, with respect to (3.43), that

$$B\lambda + C\lambda^2 = 0. \tag{3.45}$$

The roots of (3.45) are

$$\lambda_1 = 0, \quad \lambda_2 = -\frac{B}{C}. \tag{3.46}$$

Hence, (3.43) to (3.46) imply that the characteristics have equations

$$\left.\begin{array}{c} x = c_1 \\ x - \dfrac{B}{C} y = c_2. \end{array}\right\} \tag{3.47}$$

Since $\bar{x} = c_1$, $\bar{y} = c_2$ are the equations of the coordinate lines in the $\bar{x}\bar{y}$ system, the affine transformation

$$\left.\begin{array}{c} \bar{x} = x \\ \bar{y} = x - \dfrac{B}{C} y \end{array}\right\} \tag{3.48}$$

is motivated, for under this transformation the characteristics with Eqs. (3.47) map into coordinate lines in the $\bar{x}\bar{y}$ system. Mapping (3.48) implies with respect to (3.37) that $\alpha_1 = \alpha_2 = 1$, $\beta_1 = 0$, $\beta_2 = -B/C$. Then (3.29) to (3.35) imply $\bar{A} = \bar{C} = 0$, $\bar{B} = -B^2/C$, $\bar{D} = D$, $\bar{E} = D - \dfrac{BE}{C}$, $\bar{F} = F$, $\bar{G} = \bar{G}(\bar{x},\bar{y})$, so that (3.36) reduces to

$$-\frac{B^2}{C} u_{\bar{x}\bar{y}} + Du_{\bar{x}} + \left(D - \frac{BE}{C}\right)u_{\bar{y}} + Fu + \bar{G} = 0, \tag{3.49}$$

or, equivalently, to

$$u_{\bar{x}\bar{y}} = \frac{DC}{B^2} u_{\bar{x}} + \frac{(DC - BE)}{B^2} u_{\bar{y}} + \frac{\bar{F}C}{B^2} u + \frac{\bar{G}C}{B^2}. \tag{3.50}$$

Setting $D_1 = \dfrac{DC}{B^2}$, $E_1 = \dfrac{(DC - BE)}{B^2}$, $F_1 = \dfrac{\bar{F}C}{B^2}$, $G_1 = \dfrac{\bar{G}C}{B^2}$ reduces (3.50) to (3.38). Thus, the theorem is valid if exactly one of A, B, C is zero.

Finally, suppose *exactly two* of A, B, C are zero. If $B = C = 0$, then the proof for the case where none of A, B, C is zero is applicable. If $A = B = 0$, then the proof for the case $A = 0$, $B \neq 0$, $C \neq 0$ is applicable. If $A = C = 0$, $B \neq 0$, then (3.36) is easily put into form (3.38) by dividing through by B and by transposing the appropriate terms.

Since all possible cases have been considered, the theorem is proved.

Example. On E^2, the p.d.e.

$$\frac{\partial^2 u}{\partial x^2} - 5\frac{\partial^2 u}{\partial x\,\partial y} + \frac{\partial u}{\partial x} - \frac{\partial u}{\partial y} = 0 \qquad (3.51)$$

is hyperbolic since $A = 1$, $B = -5$, $C = 0$ and $B^2 - 4AC = 25 > 0$. The affine mapping

$$\bar{x} = 5x + y$$
$$\bar{y} = y$$

applied to (3.51) and transposition of terms readily yield

$$\frac{\partial^2 u}{\partial \bar{x}\,\partial \bar{y}} = \frac{4}{25}\frac{\partial u}{\partial \bar{x}} - \frac{1}{25}\frac{\partial u}{\partial \bar{y}}.$$

Theorem 3.3. Given linear p.d.e.

$$Au_{xx} + Bu_{xy} + Cu_{yy} + Du_x + Eu_y + Fu + G = 0, \qquad (3.52)$$

where A, B, C, D, E, F are real constants, G is a real-valued function of x and y on plane point set M, and $A^2 + B^2 + C^2 \neq 0$, then if (3.52) is hyperbolic on M there exists an affine transformation of the form

$$\left.\begin{array}{l} \bar{x} = \alpha_1 x + \beta_1 y \\ \bar{y} = \alpha_2 x + \beta_2 y \end{array}\right\} \qquad (3.53)$$

such that in the $\bar{x}\bar{y}$ coordinates, (3.52) has the form

$$u_{\bar{x}\bar{x}} - u_{\bar{y}\bar{y}} = D_1 u_{\bar{x}} + E_1 u_{\bar{y}} + F_1 u + G_1, \qquad (3.54)$$

where D_1, E_1, F_1 are constants and G_1 is a real-valued function of $\bar{x}$, $\bar{y}$ on the set M_1 which is the map of M under (3.53).

Proof. By Theorem 3.2, there exists an affine transformation

$$\left.\begin{array}{l} x^* = \alpha_1^* x + \beta_1^* y \\ y^* = \alpha_2^* x + \beta_2^* y \end{array}\right\} \qquad (3.55)$$

with the aid of which (3.54) can be transformed into

$$u_{x^*y^*} = D^* u_{x^*} + E^* u_{y^*} + F^* u + G^*. \qquad (3.56)$$

Then the mapping

$$\left.\begin{array}{l} \bar{x} = x^* + y^* \\ \bar{y} = x^* - y^* \end{array}\right\} \qquad (3.57)$$

applied to (3.56) readily yields (3.54).

Hence, since $(\alpha_1^* \beta_2^* - \alpha_2^* \beta_1^*) \neq 0$, one could of course combine (3.55) and (3.56) into the single affine transformation

$$\bar{x} = (\alpha_1^* + \alpha_2^*)x + (\beta_1^* + \beta_2^*)y$$
$$\bar{y} = (\alpha_1^* - \alpha_2^*)x + (\beta_1^* - \beta_2^*)y$$

with the aid of which (3.52) can be transformed into (3.54).

Example. On E^2, the p.d.e.

$$\frac{\partial^2 u}{\partial x^2} - 5 \frac{\partial^2 u}{\partial x \, \partial y} + \frac{\partial u}{\partial x} - \frac{\partial u}{\partial y} = 0$$

is hyperbolic. Under the mapping

$$\bar{x} = 5x + 2y$$
$$\bar{y} = 5x,$$

the equation transforms into

$$-25 \frac{\partial^2 u}{\partial \bar{x}^2} + 25 \frac{\partial^2 u}{\partial \bar{y}^2} + 3 \frac{\partial u}{\partial \bar{x}} + 5 \frac{\partial u}{\partial \bar{y}} = 0,$$

which readily reduces to

$$\frac{\partial^2 u}{\partial \bar{x}^2} - \frac{\partial^2 u}{\partial \bar{y}^2} = \frac{3}{25} \frac{\partial u}{\partial \bar{x}} + \frac{1}{5} \frac{\partial u}{\partial \bar{y}}.$$

Theorem 3.4. Given linear p.d.e.

$$Au_{xx} + Bu_{xy} + Cu_{yy} + Du_x + Eu_y + Fu + G = 0, \qquad (3.58)$$

where A, B, C, D, E, F are real constants, G is a real-valued function of x and y on plane point set M, and $A^2 + B^2 + C^2 \neq 0$, then if (3.58) is elliptic on M, there exists an affine transformation of the form

$$\left. \begin{array}{l} \bar{x} = \alpha_1 x + \beta_1 y \\ \bar{y} = \alpha_2 x + \beta_2 y \end{array} \right\} \qquad (3.59)$$

such that in the $\bar{x}\bar{y}$ system, (3.58) has the form

$$u_{\bar{x}\bar{x}} + u_{\bar{y}\bar{y}} = D_1 u_{\bar{x}} + E_1 u_{\bar{y}} + F_1 u + G_1, \qquad (3.60)$$

where D_1, E_1, F_1 are constants and G_1 is a real-valued function of $\bar{x}$ and $\bar{y}$ on the set M_1 which is the map of M under (3.59).

Proof. Note first that $B^2 - 4AC < 0$, so that neither A nor C is zero.

From (3.21) to (3.24), it follows that the characteristic curves for elliptic partial differential equations do not exist in the real plane; that is, since λ_1, λ_2 are complex conjugates, the characteristics are not real curves. Thus, the methods of Theorems 3.2 and 3.3 do not apply immediately. Let us reason, however, in a heuristic fashion and see if an appropriate affine transformation can, nevertheless, be found.

Let λ_1, λ_2 be the complex conjugate roots (3.24) of (3.23). Then integration of

$$\frac{dy}{dx} = -\lambda_1, \qquad \frac{dy}{dx} = -\lambda_2$$

without regard to the fact that λ_1, λ_2 are complex yields

$$y + \lambda_1 x = d_1, \qquad y + \lambda_2 x = d_2$$

for characteristic curves.   However, since λ_1, λ_2 are complex, allow d_1, d_2 also to be complex.   Moreover, setting $d_1 = c_1 + ic_2$, $d_2 = c_1 - ic_2$, where c_1, c_2 are arbitrary real constants, still allows the arbitrary choice of two real constants.   Thus

$$y + \lambda_1 x = c_1 + ic_2, \qquad y + \lambda_2 x = c_1 - ic_2. \tag{3.61}$$

If now, as in Theorem 3.2, one sets $\bar{x} = c_1$, $\bar{y} = c_2$, then (3.61) yields

$$y + \lambda_1 x = \bar{x} + i\bar{y}, \qquad y + \lambda_2 x = \bar{x} - i\bar{y}. \tag{3.62}$$

Solving (3.62) for $\bar{x}$, $\bar{y}$ and applying (3.24) yields

$$\left. \begin{aligned} \bar{x} &= y - \frac{B}{2A}\,x \\[2mm] \bar{y} &= \frac{\sqrt{4AC - B^2}}{2A}\,x. \end{aligned} \right\} \tag{3.63}$$

Application of affine transformation (3.63) to (3.58) and transposition of terms readily yields the theorem.

Example.   On E^2, the p.d.e.

$$\frac{\partial^2 u}{\partial x^2} + \frac{\partial^2 u}{\partial x\,\partial y} + \frac{\partial^2 u}{\partial y^2} - u = 0$$

is elliptic.   By means of the affine mapping

$$\bar{x} = -\tfrac{1}{2}x + y$$

$$\bar{y} = \frac{\sqrt{3}}{2}\,x$$

and transposition of terms, the p.d.e. transforms into

$$\frac{\partial^2 u}{\partial \bar{x}^2} + \frac{\partial^2 u}{\partial \bar{y}^2} = \frac{4u}{3}.$$

We shall consider finally the case where p.d.e. (3.20) is parabolic. Note first that in this case if $A = 0$, then $B^2 - 4AC = 0$ implies $B = 0$.   Thus if $A = 0$, then $A^2 + B^2 + C^2 \neq 0$ implies $C \neq 0$. Thus, the class of parabolic equations for which $A \neq 0$ and/or $C \neq 0$ includes all parabolic equations.   Therefore, only the possibilities $A \neq 0$, $C \neq 0$ need be considered.

Theorem 3.5. Given linear p.d.e.

$$Au_{xx} + Bu_{xy} + Cu_{yy} + Du_x + Eu_y + Fu + G = 0, \qquad (3.64)$$

where A, B, C, D, E, F are real constants, $A \neq 0$, $A^2 + B^2 + C^2 \neq 0$, and G is a real-valued function of x and y on plane point set M, then if (3.64) is parabolic on M, there exists an affine transformation of the form

$$\left.\begin{aligned} \bar{x} &= \alpha_1 x + \beta_1 y \\ \bar{y} &= \alpha_2 x + \beta_2 y \end{aligned}\right\} \qquad (3.65)$$

such that in the $\bar{x}\bar{y}$ coordinates (3.64) has the form

$$u_{\bar{x}\bar{x}} = D_1 u_{\bar{x}} + E_1 u_{\bar{y}} + F_1 u + G_1, \qquad (3.66)$$

where D_1, E_1, F_1 are constants, and G_1 is a real-valued function of $\bar{x}$ and $\bar{y}$ on the set M_1 which is the map of M under (3.65).

Proof. For the characteristic differential equation (3.21), (3.24) implies

$$\lambda_1 = \lambda_2 = -\frac{B}{2A}, \qquad (3.67)$$

so that only the single characteristic whose equation is

$$y - \frac{B}{2A} x = c_2 \qquad (3.68)$$

results. Thus, as in the discussion of Theorem 3.2, it is reasonable to set

$$\bar{y} = -\frac{B}{2A} x + y.$$

The only restrictive factor in the choice of α_1, β_1 for

$$\bar{x} = \alpha_1 x + \beta_1 y$$

is that the Jacobian of the transformation should not be zero; that is, α_1 and β_1 may be selected arbitrarily, provided only that

$$\alpha_1 + \frac{B}{2A} \beta_1 \neq 0. \qquad (3.69)$$

However, for $\beta_1 \neq 0$, (3.69) is equivalent to

$$\frac{\alpha_1}{\beta_1} \neq \lambda_1, \qquad \beta_1 \neq 0.$$

But, either $\lambda_1 = 0$, or $\lambda_1 \neq 0$. If $\lambda_1 = 0$, then from (3.67) $B = 0$. Since $A \neq 0$, $B = 0$, and $B^2 - 4AC = 0$, then $C = 0$. But, in this case, transposition of terms easily reduces (3.64) to form (3.66) and the theorem is proved.　Hence, assume $\lambda_1 \neq 0$.　Then one can set $\alpha_1 = -\lambda_1$, $\beta_1 = 1$, so that (3.69) is satisfied and the resulting affine transformation is

$$\left.\begin{aligned} \bar{x} &= \frac{B}{2A}\,x + y \\ \bar{y} &= -\frac{B}{2A}\,x + y. \end{aligned}\right\} \qquad (3.70)$$

Application of transformation (3.70) to (3.64) and transposition of terms readily yields the theorem.

Example.　On E^2, the equation

$$\frac{\partial^2 u}{\partial x^2} + 4\,\frac{\partial^2 u}{\partial x\,\partial y} + 4\,\frac{\partial^2 u}{\partial y^2} + \frac{\partial u}{\partial x} = 0$$

is parabolic.　By means of the affine mapping

$$\bar{x} = 2x + y$$
$$\bar{y} = -2x + y$$

and by transposition of terms, the equation transforms into

$$\frac{\partial^2 u}{\partial \bar{x}^2} = -\frac{1}{8}\frac{\partial u}{\partial \bar{x}} + \frac{1}{8}\frac{\partial u}{\partial \bar{y}}.$$

Theorem 3.6.　Given linear p.d.e. (3.64) with A, B, C, D, E, F real constants, $C \neq 0$, $A^2 + B^2 + C^2 \neq 0$, and G a real-valued function of x and y on plane point set M, then if (3.64) is parabolic on M there exists an affine transformation of form (3.65) such that in the $\bar{x}\bar{y}$ coordinates (3.64) has the form

$$u_{\bar{y}\bar{y}} = D_1 u_{\bar{x}} + E_1 u_{\bar{y}} + F_1 u + G_1, \qquad (3.71)$$

where D_1, E_1, F_1 are constants, and G_1 is a real-valued function of $\bar{x}$ and $\bar{y}$ on the set M_1 which is the map of M under (3.65).

The proof follows in a fashion analogous to that of Theorem 3.5.

Equations (3.38), (3.54), (3.60), (3.66), and (3.71) are called *canonical*, or standard, forms of linear partial differential equation (3.20).　Since (3.20), however, is always either hyperbolic, elliptic, or parabolic, it follows from Theorems 3.1 to 3.6 that a complete theory for linear, second-order partial differential equation (3.20) can be established by a study only of the canonical forms.　Finally, since (3.38) and (3.54) are equivalent and because (3.66) and (3.71) differ only in notation, it follows that a complete theory can be developed by consideration only of the three canonical forms

$$u_{xx} - u_{yy} = D_1 u_x + E_1 u_y + F_1 u + G_1, \tag{3.72}$$

$$u_{xx} + u_{yy} = D_1 u_x + E_1 u_y + F_1 u + G_1, \tag{3.73}$$

$$u_{xx} = D_1 u_x + E_1 u_y + F_1 u + G_1. \tag{3.74}$$

Guided by the applicability of (3.72) to (3.74) to the physical sciences, we proceed to a study of prototype hyperbolic, elliptic, and parabolic equations.

EXERCISES

3.1. Determine whether or not each of the following functions is a solution on E^2 of the accompanying partial differential equation:

(a) $u = x + y$; $\dfrac{\partial^2 u}{\partial x^2} + \dfrac{\partial^2 u}{\partial y^2} = 0$

(b) $u = (x + y)^7 + (x - y)^6$; $u_{xx} - u_{yy} = 0$

(c) $u = x^2 + 2y$; $\dfrac{\partial^2 u}{\partial x^2} - \dfrac{\partial u}{\partial y} = 0$

(d) $u = e^{-(x+3y)} + 2e^{x+y}\sin x$; $u_{xx} - 2u_x + u_y = 0$

(e) $u = \frac{1}{2}xy^2 + (1 + x + x^2 + x^3 + x^4 + x^5)y + e^{-x}$; $\dfrac{\partial^2 u}{\partial y^2} - x = 0$

(f) $u = e^{xy}[(y - 1)\cos xy + \sin y \cos y \sin xy] + 2y$; $u_{xx} - 2yu_x + 2y^2 u = 4y^3$

(g) $u = x^2 + y^2$; $u\dfrac{\partial^2 u}{\partial x^2} + u\dfrac{\partial^2 u}{\partial y^2} - 4x - 4y = 0$

(h) $u = f(x + y) + g(x - y)$; f, g arbitrary functions of class C^2; $u_{xx} - u_{yy} = 0$

(i) $u = f(2x + y) + g(x - y) - xy$;
f, g arbitrary functions of class C^2; $\dfrac{\partial^2 u}{\partial x^2} - \dfrac{\partial^2 u}{\partial x\,\partial y} - 2\dfrac{\partial^2 u}{\partial y^2} = 1$

(j) $u = f(x + y) + g(2x + y)$;
f, g arbitrary functions of class C^2; $u_{xx} - 3u_{xy} + 2u_{yy} = 0$.

3.2. Let $(x,y) \in M$ if and only if $x > 0$, $y > 0$. Show that each of the following functions is a solution of the accompanying partial differential equation on M:

(a) $u = \log(x^2 + y^2)$; $\dfrac{\partial^2 u}{\partial x^2} + \dfrac{\partial^2 u}{\partial y^2} = 0$

(b) $u = \dfrac{y}{x}$; $u_{yy} - yu_{xy} + u_x = 0$.

3.3. Let $(s,t) \in M$ if and only if both s and t are real. Show that each of the following functions is a solution of the accompanying partial differential equation on M:

(a) $u = s^2 - t^2$; $\qquad \dfrac{\partial^2 u}{\partial s^2} + \dfrac{\partial^2 u}{\partial t^2} = 0$

(b) $u = \sin s \cos t$; $\qquad \dfrac{\partial^2 u}{\partial s^2} - \dfrac{\partial^2 u}{\partial t^2} + u\dfrac{\partial u}{\partial s} - \cos s \sin t\dfrac{\partial u}{\partial t} = \tfrac{1}{2}\sin 2s.$

3.4. Sketch each of the following space curves and give direction numbers of the tangent vector at each point (x,y,u) of the curve:

(a) $x = t, y = t^2, u = t^3$; $\qquad t \geq 0$
(b) $x = \sin t, y = \cos t, u = t$; $\qquad t \geq 0$
(c) $x = \sin t, y = \cos t, u = t^2$; $\qquad t \geq 0$
(d) $x = t, y = t^2, u = e^t$; $\qquad t \geq 0.$

3.5. Sketch in the xy plane the projections of the curves of Exercise 3.4.

3.6. Find direction numbers of the normal and the equation of the tangent plane at the point $(1,1,2)$ for the surface whose equation is

(a) $u = x^2 + y^2$ $\qquad\qquad$ (Ans. $2x + 2y - u = 2$)
(b) $u = x^2 + y$ $\qquad\qquad$ (Ans. $2x + y - u = 1$)
(c) $x^2 + y^2 + u^2 = 6$ $\qquad\qquad$ (Ans. $x + y + 2u = 6$)
(d) $2x^2 + y^2 - u = 1$
(e) $x^2 + y^2 = 2$
(f) $x - y - u + 2 = 0.$

3.7. Let M be the set of real numbers. Show that, on $E^2, u = x + \tfrac{1}{6}[(x+y)^3 - (x-y)^3]$ is a solution while $u = x - y$ is not a solution of the initial-value problem for partial differential equation

$$u_{xx} - u_{yy} = 0$$

with C_0: $x = t, y = 0, t \in M$, and with initial conditions

(a) $f(t,0) \equiv t, \qquad t \in M$

(b) $\dfrac{\partial f(t,0)}{\partial x} \equiv 1, \dfrac{\partial f(t,0)}{\partial y} \equiv t^2, \qquad t \in M.$

3.8. Let M be the set of real numbers. Show that on $E^2, u = \tfrac{1}{2}[(x+2y)^2 + (x-2y)^2] + \tfrac{1}{4}[(x+2y)^4 - (x-2y)^4]$ is a solution, while $u = x^2 - y^2$ is not a solution of the initial-value problem for partial differential equation

$$4\dfrac{\partial^2 u}{\partial x^2} - \dfrac{\partial^2 u}{\partial y^2} = 0$$

with C_0: $x = t^2, y = 0; t \in M$, and with initial conditions

(a) $f(t^2,0) \equiv t^2, \qquad t \in M$

(b) $\dfrac{\partial f(t^2,0)}{\partial x} \equiv 2t^2, \dfrac{\partial f(t^2,0)}{\partial y} \equiv 4t^6, \qquad t \in M.$

3.9. Find the equations of the characteristics of each of the following partial differential equations:

(a) $\dfrac{\partial^2 u}{\partial x^2} + 2\dfrac{\partial^2 u}{\partial x\,\partial y} + \dfrac{\partial^2 u}{\partial y^2} = 0$ $\qquad\qquad$ (Ans. $y = x + c$)

(b) $\dfrac{\partial^2 u}{\partial x^2} + 3\,\dfrac{\partial^2 u}{\partial x\,\partial y} + 2\,\dfrac{\partial^2 u}{\partial y^2} = 0$ $\qquad\qquad$ (*Ans.* $y = 2x + c_1,\, y = x + c_2$)

(c) $\dfrac{\partial^2 u}{\partial x^2} + 4\,\dfrac{\partial^2 u}{\partial x\,\partial y} + 5\,\dfrac{\partial^2 u}{\partial y^2} = 0$ $\qquad\qquad$ (*Ans.* No real characteristics)

(d) $\dfrac{\partial^2 u}{\partial x^2} - 2\,\dfrac{\partial^2 u}{\partial x\,\partial y} + \dfrac{\partial^2 u}{\partial y^2} = 0$

(e) $\dfrac{\partial^2 u}{\partial x^2} - 3\,\dfrac{\partial^2 u}{\partial x\,\partial y} + \dfrac{\partial^2 u}{\partial y^2} - \dfrac{\partial u}{\partial x} = 0$

(f) $2\,\dfrac{\partial^2 u}{\partial x\,\partial y} + \dfrac{\partial^2 u}{\partial y^2} - 2\,\dfrac{\partial u}{\partial x} - 3\,\dfrac{\partial u}{\partial y} + 4u + e^{xy} = 0.$

3.10. Classify each of the following partial differential equations as hyperbolic, elliptic, or parabolic on E^2:

(a) $u_{xx} + u_{xy} + u_{yy} = 0$ $\qquad\qquad\qquad\qquad\qquad$ (*Ans.* Elliptic)
(b) $u_{xx} + 2u_{xy} + u_{yy} = 0$ $\qquad\qquad\qquad\qquad\quad$ (*Ans.* Parabolic)
(c) $u_{xx} + u_{xy} - u_{yy} = 0$ $\qquad\qquad\qquad\qquad\quad$ (*Ans.* Hyperbolic)
(d) $u_{xx} + 4u_{xy} + u_{yy} = 0$
(e) $u_{xx} - u_{xy} - u_{yy} = 0$
(f) $u_{xx} - 2u_{xy} + u_{yy} = 0$
(g) $u_{xx} - 2u_{xy} + u_{yy} + u_y = 0$
(h) $u_{xx} + u_{xy} + 3u_{yy} - 7 = 0$
(i) $u_{xx} - 3u_{xy} + u_{yy} - u_x = 0$
(j) $2u_{xx} - u_{xy} + u_x - u_y + u = 1$
(k) $2u_{xy} + u_{yy} + 2u_x - 3u_y + 4u + e^{xy} = 0$
(l) $u_{xx} - u_{xy} + 3u_x - 4u_y + u = x^2 + y^2.$

3.11. Transform the following partial differential equations, which are hyperbolic on E^2, into canonical forms (3.38) and (3.54):

(a) $\dfrac{\partial^2 u}{\partial x^2} + \dfrac{\partial^2 u}{\partial x\,\partial y} - \dfrac{\partial^2 u}{\partial y^2} = 0$

(b) $\dfrac{\partial^2 u}{\partial x^2} + 4\,\dfrac{\partial^2 u}{\partial x\,\partial y} + \dfrac{\partial^2 u}{\partial y^2} = 0$

(c) $\dfrac{\partial^2 u}{\partial x^2} - \dfrac{\partial^2 u}{\partial x\,\partial y} - \dfrac{\partial^2 u}{\partial y^2} = 0$

(d) $\dfrac{\partial^2 u}{\partial x^2} + \dfrac{\partial^2 u}{\partial x\,\partial y} - 3\,\dfrac{\partial^2 u}{\partial y^2} + 7 = 0$

(e) $\dfrac{\partial^2 u}{\partial x^2} - 3\,\dfrac{\partial^2 u}{\partial x\,\partial y} + \dfrac{\partial^2 u}{\partial y^2} - \dfrac{\partial u}{\partial x} = 0$

(f) $\dfrac{\partial^2 u}{\partial x^2} + \dfrac{\partial^2 u}{\partial x\,\partial y} - 3\,\dfrac{\partial^2 u}{\partial y^2} + 7\,\dfrac{\partial u}{\partial x} - 8\,\dfrac{\partial u}{\partial y} + e^{xy} = 0.$

3.12. Transform the following partial differential equations, which are elliptic on E^2, into canonical form (3.60):

(a) $u_{xx} + u_{xy} + u_{yy} = 0$
(b) $u_{xx} + u_{xy} + 3u_{yy} = 0$
(c) $2u_{xx} - u_{xy} + u_{yy} = 0$
(d) $u_{xx} - 2u_{xy} + 5u_{yy} + u_y = 0$
(e) $2u_{xx} + 3u_{xy} + 4u_{yy} + u_x - e^{xy} = 1$.

3.13. Transform the following partial differential equations, which are parabolic on E^2, into either canonical form (3.66) or canonical form (3.71):

(a) $\dfrac{\partial^2 u}{\partial x^2} + 2\dfrac{\partial^2 u}{\partial x\,\partial y} + \dfrac{\partial^2 u}{\partial y^2} = 0$

(b) $\dfrac{\partial^2 u}{\partial x^2} - 2\dfrac{\partial^2 u}{\partial x\,\partial y} + \dfrac{\partial^2 u}{\partial y^2} = 0$

(c) $\dfrac{\partial^2 u}{\partial x^2} + 2\dfrac{\partial^2 u}{\partial x\,\partial y} + \dfrac{\partial^2 u}{\partial y^2} + 7\dfrac{\partial u}{\partial x} - 8\dfrac{\partial u}{\partial y} = 0$

(d) $\dfrac{\partial^2 u}{\partial x^2} - 4\dfrac{\partial^2 u}{\partial x\,\partial y} + 4\dfrac{\partial^2 u}{\partial y^2} - \dfrac{\partial u}{\partial x} - x^2 = 0$

(e) $\dfrac{\partial^2 u}{\partial x^2} - 6\dfrac{\partial^2 u}{\partial x\,\partial y} + 9\dfrac{\partial^2 u}{\partial y^2} + \dfrac{\partial u}{\partial x} - e^{xy} = 1$.

3.14. By selecting appropriate exponential transformations show that on E^2
(a) canonical form (3.72) can be further simplified to the form

$$u_{\xi\xi} - u_{\eta\eta} = F^*u + G^*(\xi,\eta)$$

(b) canonical form (3.73) can be further simplified to the form

$$u_{\xi\xi} + u_{\eta\eta} = F^*u + G^*(\xi,\eta)$$

(c) canonical form (3.74) can be further simplified to the form

$$u_{\xi\xi} = u_\eta + G^*(\xi,\eta)$$

3.15. A quasi-linear partial differential equation

$$Au_{xx} + Bu_{xy} + Cu_{yy} = \Phi(x,y,u,u_x,u_y),$$

where $A = A(x,y)$, $B = B(x,y)$, $C = C(x,y)$ and $A^2 + B^2 + C^2$ is never zero, is said to be:
(a) hyperbolic at those points where $B^2 - 4AC > 0$
(b) elliptic at those points where $B^2 - 4AC < 0$
(c) parabolic at those points where $B^2 - 4AC = 0$.
Show that

$$u_{xx} + 2xu_{xy} + (1 - y^2)u_{yy} = 0$$

is
(a) hyperbolic for all $(x,y) \in E^2$ whose coordinates satisfy $x^2 + y^2 > 1$
(b) elliptic for all $(x,y) \in E^2$ whose coordinates satisfy $x^2 + y^2 < 1$
(c) parabolic for all $(x,y) \in E^2$ whose coordinates satisfy $x^2 + y^2 = 1$.

Chapter 4

THE WAVE EQUATION

Investigations in the theory of wave propagation lead quite naturally to the consideration of a linear, hyperbolic partial differential equation called the wave equation.

Definition 4.1. The equation

$$\frac{\partial^2 u}{\partial x^2} - \frac{\partial^2 u}{\partial y^2} = G(x,y), \tag{4.1}$$

where $G(x,y)$ is defined on plane point set M, is called the wave equation, or, more precisely, the one-dimensional wave equation.

Definition 4.2. On E^2, the equation

$$\frac{\partial^2 u}{\partial x^2} - \frac{\partial^2 u}{\partial y^2} = 0 \tag{4.2}$$

is called the homogeneous wave equation.

4.1 The Homogeneous Wave Equation

Since (4.2) is, perhaps, the simplest form of p.d.e. (4.1), we consider it first. The affine mapping

$$\bar{x} = x + y$$
$$\bar{y} = x - y$$

transforms (4.2) into

$$4 \frac{\partial^2 u}{\partial \bar{x} \, \partial \bar{y}} = 0,$$

which is equivalent to

$$\frac{\partial^2 u}{\partial \bar{x} \, \partial \bar{y}} = 0. \tag{4.3}$$

In general, then, p.d.e. (4.3) implies that

$$\frac{\partial u}{\partial \bar{x}} = \phi(\bar{x}),$$

which, in turn, implies that

$$u = \int_0^{\bar{x}} \phi(t)\, dt + f_2(\bar{y})$$
$$= f_1(\bar{x}) + f_2(\bar{y})$$
$$= f_1(x + y) + f_2(x - y).$$

Thus, the following theorem is motivated.

Theorem 4.1. Let N_1, N_2 be nonempty sets of real numbers. Define M_1 as all points (x,y) in E^2 whose coordinates satisfy $x + y = t_1$, $t_1 \in N_1$ and define M_2 as all points (x,y) in E^2 whose coordinates satisfy $x - y = t_2$, $t_2 \in N_2$. Set $M = M_1 \cap M_2$. If $f_1(t_1) \in C^2$ on N_1, $f_2(t_2) \in C^2$ on N_2, then

$$f(x,y) = f_1(x + y) + f_2(x - y) \qquad (4.4)$$

is a solution of the homogeneous wave equation (4.2) on M. Conversely, if $f(x,y)$ is a solution of the homogeneous wave equation (4.2) on M, then there exist functions $f_1(t_1) \in C^2$ on N_1, $f_2(t_2) \in C^2$ on N_2 such that

$$f(x,y) = f_1(x + y) + f_2(x - y).$$

Proof. The proof of the first part of the theorem follows by differentiation and application of Definition 3.4. The proof of the converse follows from the discussion which preceded the statement of the theorem.

Example 1. $f(x,y) = (x + y)^3 - \sin(x - y)$ is a solution on E^2 of p.d.e. (4.2).

Example 2. $f(x,y) = 4(x + y)^5 \sin(x + y) - (x - y)^3 e^{x-y}$ is a solution on E^2 of p.d.e. (4.2).

With the aid of Theorem 4.1, the following very basic result can be established.

Theorem 4.2. In E^2, consider a rectangle whose sides are segments of characteristics of the homogeneous wave equation. Label the vertices P_1, P_2, P_3, P_4, as in Diagram 4.1, with P_1, P_4 as opposite vertices. Also if $v(x,y)$ is a function defined on E^2, then the notation $v(P)$ will designate the value of $v(x,y)$ at the point P. Then a necessary and sufficient condition that a function $u = f(x,y)$, which is of class C^2 on E^2, be a solution of homogeneous wave equation (4.2) on E^2 is that for all possible rectangles, as described above, $f(x,y)$ satisfy the difference equation

$$f(P_1) + f(P_4) = f(P_2) + f(P_3). \tag{4.5}$$

Proof. Let $u = f(x,y)$ be a solution of class C^2 on E^2 of p.d.e. (4.2). Then if P_4 is assigned coordinates (x_1,y_1), one may assign coordinates $(x_1 - c, y_1 + c)$, $(x_1 + d, y_1 + d)$, $(x_1 + d - c, y_1 + d + c)$, respectively, to P_2, P_3, P_1. Then set $c = h - k$, $d = h + k$, so that

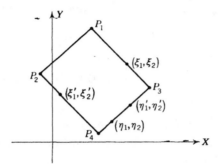

DIAGRAM 4.1

the coordinates of P_1, P_2, P_3, P_4 become $(x_1 + 2k, y_1 + 2h)$, $(x_1 - h + k, y_1 + h - k)$, $(x_1 + h + k, y_1 + h + k)$, (x_1,y_1), respectively. Finally, setting $x = x_1 + k$, $y = y_1 + h$, the coordinates of P_1, P_2, P_3, P_4 reduce to $(x + k, \ y + h)$, $(x - h, \ y - k)$, $(x + h, y + k)$, $(x - k, y - h)$, respectively.

From (4.4) then

$$\begin{aligned}
f(P_1) + f(P_4) &= f_1(x + k + y + h) + f_2(x + k - y - h) \\
&\quad + f_1(x - k + y - h) + f_2(x - k - y + h) \\
&= f_1(x - h + y - k) + f_2(x - h - y + k) \\
&\quad + f_1(x + h + y + k) + f_2(x + h - y - k) \\
&= f(P_2) + f(P_3),
\end{aligned}$$

which completes the proof of the necessity.

For the sufficiency, let P_1, P_2, P_3, P_4 have coordinates $(x_1 + d - c, y_1 + d + c)$, $(x_1 - c, y_1 + c)$, $(x_1 + d, y_1 + d)$, (x_1,y_1), respectively, and let (4.5) be valid. The discussion will be limited to the case $c > 0$, $d > 0$, since the other cases follow in a similar fashion.

It follows from (4.5) that

$$f(P_1) - f(P_3) = f(P_2) - f(P_4). \tag{4.6}$$

Hence,

$$f(x_1 + d - c, y_1 + d + c) - f(x_1 + d, y_1 + d)$$
$$= f(x_1 - c, y_1 + c) - f(x_1, y_1). \quad (4.7)$$

Application of the mean-value theorem to (4.7) yields

$$-cf_x(\xi_1, \xi_2) + cf_y(\xi_1, \xi_2) = -cf_x(\xi_1', \xi_2') + cf_y(\xi_1', \xi_2'), \quad (4.8)$$

where (ξ_1, ξ_2), (ξ_1', ξ_2') are located as shown in Diagram 4.1. However, the assumption $c > 0$ implies with respect to (4.8) that

$$-f_x(\xi_1, \xi_2) + f_y(\xi_1, \xi_2) = -f_x(\xi_1', \xi_2') + f_y(\xi_1', \xi_2'). \quad (4.9)$$

Hence, as $c \to 0$, (4.9) implies

$$-f_x(x_1 + d, y_1 + d) + f_y(x_1 + d, y_1 + d) = -f_x(x_1, y_1) + f_y(x_1, y_1), \quad (4.10)$$

or, equivalently, that

$$f_x(x_1 + d, y_1 + d) - f_x(x_1, y_1) = f_y(x_1 + d, y_1 + d) - f_y(x_1, y_1). \quad (4.11)$$

Application of the mean-value theorem to (4.11) yields

$$d[f_{xx}(\eta_1, \eta_2)] + d[f_{xy}(\eta_1, \eta_2)] = d[f_{xy}(\eta_1', \eta_2')] + d[f_{yy}(\eta_1', \eta_2')], \quad (4.12)$$

where (η_1, η_2), (η_1', η_2') are located on segment $P_3 P_4$. Since $d > 0$, (4.12) implies

$$f_{xx}(\eta_1, \eta_2) + f_{xy}(\eta_1, \eta_2) = f_{xy}(\eta_1', \eta_2') + f_{yy}(\eta_1', \eta_2'). \quad (4.13)$$

Hence, as $d \to 0$, (4.13) then implies

$$f_{xx}(x_1, y_1) = f_{yy}(x_1, y_1). \quad (4.14)$$

Since (x_1, y_1) is an arbitrary point of E^2, then, from (4.14),

$$f_{xx}(x, y) - f_{yy}(x, y) \equiv 0$$

on E^2, and the proof is complete.

Note that the proof of Theorem 4.2 holds on point sets other than E^2 so that, for example, the theorem, appropriately restated, is valid also on a semi-infinite strip $\bar{R}$ defined as follows. Let a, b be two real numbers with the property $a < b$. Then $(x, y) \in \bar{R}$ if and only if $a \le x \le b$, $y \ge 0$.

4.2 The First Cauchy Problem

Solution (4.4) of the homogeneous wave equation is exceedingly general. Further conditions, in addition to that of satisfying partial differential equation (4.2) on some plane point set, must therefore be imposed if one wishes to consider a problem which is to have a unique solution. Depending on the types of additional conditions to be imposed, the problems to be explored will be called *Cauchy problems* and *mixed-type problems*. Each has physical significance and will be considered now in turn.

The first problem is described as follows.

Cauchy Problem I. Let N be the set of all real numbers. Find a solution $u = f(x,y)$ on E^2 of the homogeneous wave equation such that if $p(x) \in C^2$, $q(x) \in C^1$, $x \in N$, then

$$f(x,0) \equiv p(x), \qquad x \in N \qquad (4.15)$$

$$\frac{\partial f(x,0)}{\partial y} \equiv q(x), \qquad x \in N. \qquad (4.16)$$

Essentially, Cauchy Problem I requires one to find a solution of the homogeneous wave equation which assumes a prescribed function value and a prescribed normal derivative at each point of the x axis. Also, if one were to denote the x axis by C_0 and note that the given values $f(x,0)$ determine $\partial f/\partial x$ along the x axis, then it is readily seen that Cauchy Problem I is an initial-value problem (see Definition 3.5).

Setting

$$\left. \begin{array}{l} \bar{x} = x + y \\ \bar{y} = x - y \end{array} \right\} \qquad (4.17)$$

in solution (4.4) yields

$$f(x,y) = f_1(\bar{x}) + f_2(\bar{y}). \qquad (4.18)$$

For $y = 0$, (4.15), (4.17), and (4.18) imply

$$f_1(x) + f_2(x) = p(x), \qquad x \in N. \qquad (4.19)$$

Moreover, from (4.17) and (4.18),

$$\frac{\partial f(x,y)}{\partial y} \equiv \frac{\partial f_1(\bar{x})}{\partial \bar{x}} \frac{\partial \bar{x}}{\partial y} + \frac{\partial f_2(\bar{y})}{\partial \bar{y}} \frac{\partial \bar{y}}{\partial y} \equiv \frac{\partial f_1(\bar{x})}{\partial \bar{x}} - \frac{\partial f_2(\bar{y})}{\partial \bar{y}}, \qquad (4.20)$$

so that, for $y = 0$, (4.16), (4.17), and (4.20) imply

$$\frac{\partial f_1(x)}{\partial x} - \frac{\partial f_2(x)}{\partial x} = q(x), \qquad x \in N. \qquad (4.21)$$

Rewriting (4.21) as

$$f_1'(x) - f_2'(x) = q(x), \qquad x \in N \qquad (4.22)$$

and differentiating (4.19) with respect to x to give

$$f_1'(x) + f_2'(x) = p'(x) \qquad (4.23)$$

yields a pair of linear, algebraic equations (4.22) and (4.23) in f_1', f_2'.

The solution of this system is

$$f_1'(x) = \frac{p'(x) + q(x)}{2}, \qquad f_2'(x) = \frac{p'(x) - q(x)}{2}, \qquad x \in N, \quad (4.24)$$

which implies

$$f_1(x) = \frac{1}{2}\left[p(x) + \int_0^x q(t)\, dt \right] + c_1,$$
$$\qquad\qquad\qquad\qquad\qquad\qquad x \in N. \qquad (4.25)$$
$$f_2(x) = \frac{1}{2}\left[p(x) - \int_0^x q(t)\, dt \right] + c_2,$$

Then (4.17), (4.18), and (4.25) imply

$$f(x,y) = \frac{p(x + y) + p(x - y)}{2} + \frac{1}{2} \int_{x-y}^{x+y} q(t)\, dt + c_1 + c_2. \quad (4.26)$$

From condition (4.15), however, it is necessary that $c_1 + c_2 = 0$, so that one finally arrives at the following theorem.

Theorem 4.3. Let N be the set of real numbers. If $p(x) \in C^2$, $q(x) \in C^1$, $x \in N$, then

$$f(x,y) = \frac{p(x + y) + p(x - y)}{2} + \frac{1}{2} \int_{x-y}^{x+y} q(t)\, dt \qquad (4.27)$$

is the unique solution on E^2 of the homogeneous wave equation which satisfies conditions (4.15) and (4.16).

Proof. The proof follows directly from the above discussion and from the direct verification that (4.27) is a solution with the desired properties.

The relationship (4.27) is called the formula of D'Alembert. Note that it is of form (4.4).

Example. Find the solution $u = f(x,y)$ on E^2 of the Cauchy problem

$$u_{xx} - u_{yy} = 0, \qquad f(x,0) = x^2, \qquad \frac{\partial f(x,0)}{\partial y} = 4x^3.$$

Solution. Since $p(x) = x^2$, $q(x) = 4x^3$, the D'Alembert formula yields

$$f(x,y) = \frac{(x + y)^2 + (x - y)^2 + (x + y)^4 - (x - y)^4}{2}.$$

Corollary. If $u = f(x, y)$ is the unique solution on E^2 of Cauchy Problem I, then f depends continuously on p and q.

Proof. The proof follows directly from the formula of D'Alembert.

Various other qualitative factors concerning the unique solution of Cauchy Problem I are clearly revealed by the D'Alembert formula

(4.27). If, for example, $p(x)$ does not belong to class C^3, then $f(x,y)$ need not belong to class C^3. Thus, a solution of the homogeneous wave equation need not be an analytic function at any point of the plane. Also, suppose that (x_1,y_1) is a fixed point of the plane. The characteristics of p.d.e. (4.2) through (x_1,y_1) intersect the x axis in the two points $(x_1 - y_1, 0)$, $(x_1 + y_1, 0)$ (see Diagram 4.2), and from (4.27) it follows that $f(x_1,y_1)$ is determined completely

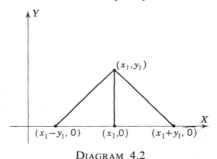

DIAGRAM 4.2

in terms of the points on the x axis which satisfy

$$x_1 - y_1 \leq x \leq x_1 + y_1.$$

Definition 4.3. Given point (x_1,y_1), then those values of x which satisfy $x_1 - y_1 \leq x \leq x_1 + y_1$ constitute the *domain of dependence* with respect to (x_1,y_1) for Cauchy Problem I.

Example. For the Cauchy problem

$$u_{xx} - u_{yy} = 0, \qquad f(x,0) = x^2, \qquad \frac{\partial f(x,0)}{\partial y} = 4x^3,$$

the domain of dependence with respect to the point $(-3,5)$ is $-8 \leq x \leq 2$.

4.3   The Second Cauchy Problem

Consider next a Cauchy problem for the nonhomogeneous wave equation (4.1).

Cauchy Problem II. Let $G(x,y) \in C^2$ on E^2 and let N be the set of all real numbers. Then find a solution $u = f(x,y)$ of wave equation (4.1) such that if $p(x) \in C^2$, $q(x) \in C^1$, $x \in N$, then

$$f(x,0) = p(x), \qquad x \in N \qquad\qquad (4.28)$$

$$\frac{\partial f(x,0)}{\partial y} = q(x), \qquad x \in N. \qquad\qquad (4.29)$$

Anticipating that the ideas developed for Cauchy Problem I will extend to Cauchy Problem II, one might reason as follows.

Let (x_1, y_1) be a point of the plane. Then the characteristics of p.d.e. (4.1) meet the x axis in the points $(x_1 - y_1, 0)$, $(x_1 + y_1, 0)$. Let the triangle with vertices (x_1, y_1), $(x_1 - y_1, 0)$, $(x_1 + y_1, 0)$ have its sides designated by C_0, C_1, C_2, as indicated in Diagram 4.3. Also,

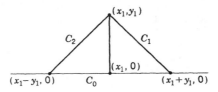

DIAGRAM 4.3

let R represent the interior of the triangle and let $\bar{R}$ equal $R \cup C_0 \cup C_1 \cup C_2$. Then, if $G(x,y)$ is integrable, p.d.e. (4.1) implies

$$\iint_{\bar{R}} (u_{xx} - u_{yy}) \, d\bar{R} = \iint_{\bar{R}} G(x,y) \, d\bar{R}. \qquad (4.30)$$

However, by Gauss's theorem, it follows that

$$\iint_{\bar{R}} (u_{xx} - u_{yy}) \, d\bar{R} \equiv \oint_{C} (u_x \, dy + u_y \, dx), \qquad (4.31)$$

where $C = C_0 \cup C_1 \cup C_2$ and C is oriented in a counterclockwise direction. The orientation of C thus induces an orientation upon C_0, C_1, C_2 with respect to which one has

$$\int_{C_0} (u_x \, dy + u_y \, dx) = \int_{x_1 - y_1}^{x_1 + y_1} u_y \, dx,$$

$$\int_{C_1} (u_x \, dy + u_y \, dx) = \int_{C_1} (-u_x \, dx - u_y \, dy)$$

$$= \int_{u(x_1 + y_1, 0)}^{u(x_1, y_1)} (-1) \, du = u(x_1 + y_1, 0) - u(x_1, y_1),$$

$$\int_{C_2} (u_x \, dy + u_y \, dx) = \int_{C_2} (u_x \, dx + u_y \, dy)$$

$$= \int_{u(x_1, y_1)}^{u(x_1 - y_1, 0)} du = u(x_1 - y_1, 0) - u(x_1, y_1).$$

Thus

$$\oint_{C} (u_x \, dy + u_y \, dx) = -2u(x_1, y_1) + u(x_1 - y_1, 0)$$

$$+ u(x_1 + y_1, 0) + \int_{x_1 - y_1}^{x_1 + y_1} u_y \, dx, \qquad (4.32)$$

so that (4.30) to (4.32) imply

$$u(x_1,y_1) = \frac{u(x_1 + y_1, 0) + u(x_1 - y_1, 0)}{2}$$
$$+ \frac{1}{2} \int_{x_1-y_1}^{x_1+y_1} u_y \, dx - \frac{1}{2} \iint_{\bar{R}} G(x,y) \, d\bar{R}. \quad (4.33)$$

If then $u = f(x,y)$ is to be a solution of Cauchy Problem II, one deduces from (4.28), (4.29), and (4.33) that

$$f(x_1,y_1) = \frac{p(x_1 + y_1) + p(x_1 - y_1)}{2} + \frac{1}{2} \int_{x_1-y_1}^{x_1+y_1} q(x) \, dx - \frac{1}{2} \iint_{\bar{R}} G(x,y) \, d\bar{R}.$$

Thus the following theorem is motivated.

Theorem 4.4. Let N be the set of all real numbers, let $p(x) \in C^2$, $q(x) \in C^1$, $x \in N$, and let $G(x,y) \in C^2$ on E^2. Then the unique solution $u = f(x,y)$ of Cauchy Problem II is given at any point (x_1,y_1) by

$$f(x_1,y_1) = \frac{p(x_1 + y_1) + p(x_1 - y_1)}{2}$$
$$+ \frac{1}{2} \int_{x_1-y_1}^{x_1+y_1} q(x) \, dx - \frac{1}{2} \iint_{\bar{R}} G(x,y) \, d\bar{R}, \quad (4.34)$$

where $\bar{R}$ is the union of the interior and the boundary of the triangle with vertices (x_1,y_1), $(x_1 - y_1, 0)$, $(x_1 + y_1, 0)$.

Proof. The proof follows from the discussion above and by direct verification that (4.34) is a solution with the desired properties.

Example 1. Find the value of the solution $u = f(x,y)$ at the point $(1,1)$ of the Cauchy problem

$$u_{xx} - u_{yy} = 1, \quad f(x,0) = x^2, \quad f_y(x,0) = 1.$$

Solution. From (4.34)

$$f(1,1) = \frac{p(2) + p(0)}{2} + \frac{1}{2} \int_0^2 q(x) \, dx - \frac{1}{2} \iint_{R} G(x,y) \, d\bar{R}$$

$$= \frac{2^2 + 0^2}{2} + \frac{1}{2} \int_0^2 1 \, dx - \frac{1}{2} \int_0^1 \int_y^{2-y} 1 \, dx \, dy$$

$$= \frac{5}{2}.$$

Example 2. Find the solution $u = f(x,y)$ at the point (x,y) of the Cauchy problem

$$u_{xx} - u_{yy} = 1, \quad f(x,0) = x^2, \quad f_y(x,0) = 1.$$

Solution. Consider (x_1, y_1) to be a point of the plane. Then

$$f(x_1, y_1) = \frac{(x_1 + y_1)^2 + (x_1 - y_1)^2}{2} + \frac{1}{2} \int_{x_1-y_1}^{x_1+y_1} dx - \frac{1}{2} \int_0^{y_1} \int_{y+x_1-y_1}^{-y+x_1+y_1} dx \, dy$$

$$= x_1^2 + \tfrac{1}{2} y_1^2 + y_1.$$

Finally, letting $(x_1, y_1) = (x, y)$ yields the solution

$$f(x, y) = x^2 + \tfrac{1}{2} y^2 + y.$$

Corollary. If $u = f(x, y)$ is the unique solution on E^2 of Cauchy Problem II, then f depends continuously on p, q, and G.

Proof. The proof follows directly from (4.34).

4.4 The Mixed-type Problem

Consider next a problem which is different from a Cauchy problem.

The Mixed-type Problem. Let a, b be two real numbers with $a < b$. Denote by R the set of plane points (x, y) whose coordinates satisfy $a < x < b$, $y > 0$, and denote by $\bar{R}$ the set of plane points (x, y) whose coordinates satisfy $a \leq x \leq b$, $y \geq 0$. Then find a function $u = f(x, y)$ on $\bar{R}$ which is a solution of the homogeneous wave equation (4.2) on R such that if $g_1(x)$, $g_2(x)$ are functions of x defined on $a \leq x \leq b$, and $m_1(y)$, $m_2(y)$ are functions of y defined for all $y \geq 0$, then

$$f(x, 0) = g_1(x), \qquad a \leq x \leq b, \tag{4.35}$$

$$\frac{\partial f(x, 0)}{\partial y} = g_2(x), \qquad a \leq x \leq b, \tag{4.36}$$

$$f(a, y) = m_1(y), \qquad y \geq 0, \tag{4.37}$$

and

$$f(b, y) = m_2(y), \qquad y \geq 0. \tag{4.38}$$

Note that mixed-type problems are also called mixed boundary-value problems and that no assumptions have been imposed germane to continuity or differentiability of $g_1(x)$, $g_2(x)$, $m_1(y)$, $m_2(y)$.

Geometrically, the mixed-type problem requires that one find a solution of the homogeneous wave equation on a semi-infinite strip such that the solution assumes prescribed function values on the sides of the strip and prescribed function and normal derivative values on the base.

Theorem 4.5. If a mixed-type problem has at least one solution $u = f(x, y)$ which is of class C^2 on $\bar{R}$, then that solution is the unique such solution.

Proof. Let $f_1(x,y)$, $f_2(x,y)$ be two solutions of class C^2 on $\bar{R}$ of a given mixed-type problem. Then $f(x,y) = f_1(x,y) - f_2(x,y)$ is a solution of the homogeneous wave equation on R which satisfies $f(x,0) = \dfrac{\partial f(x,0)}{\partial y} = 0$, $a \leq x \leq b$; $f(a,y) = f(b,y) = 0$, $y \geq 0$. It suffices therefore to show that the unique solution of class C^2 on $\bar{R}$ of the mixed-type problem

$$\left.\begin{array}{ll} u_{xx} - u_{yy} = 0, & \text{on } R \\[2mm] f(x,0) = \dfrac{\partial f(x,0)}{\partial y} = 0, & a \leq x \leq b \\[2mm] f(a,y) = f(b,y) = 0, & y \geq 0 \end{array}\right\} \tag{4.39}$$

is $f(x,y) \equiv 0$.

Consider then the integral

$$I(y) = \frac{1}{2} \int_a^b [(u_x)^2 + (u_y)^2]\, dx. \tag{4.40}$$

Then

$$\begin{aligned}
\frac{dI(y)}{dy} &= \int_a^b [u_x u_{xy} + u_y u_{yy}]\, dx \\
&= \int_a^b [u_x u_{xy} + u_y u_{xx}]\, dx \\
&= \int_a^b \frac{\partial}{\partial x} [u_x u_y]\, dx \\
&= u_x(b,y)u_y(b,y) - u_x(a,y)u_y(a,y) \\
&= u_x(b,y) \cdot 0 - u_x(a,y) \cdot 0 \\
&= 0.
\end{aligned}$$

Hence,

$$I(y) = \text{constant}. \tag{4.41}$$

However,

$$\begin{aligned}
I(0) &= \frac{1}{2} \int_a^b \{[u_x(x,0)]^2 + [u_y(x,0)]^2\}\, dx \\
&= \frac{1}{2} \int_a^b (0 + 0)\, dx \\
&= 0,
\end{aligned}$$

so that

$$I(y) \equiv 0. \tag{4.42}$$

However, (4.40) and (4.42) imply

$$(u_x)^2 + (u_y)^2 \equiv 0,$$

from which it follows that

$$u_x \equiv 0, \quad u_y \equiv 0. \tag{4.43}$$

But (4.43) yields

$$u \equiv \text{constant},$$

and since $f(x,0) \equiv 0$, it results that

$$u = f(x,y) \equiv 0.$$

Finally, since $f(x,y) \equiv 0$ is a solution of (4.39), the theorem is proved.

4.5   The Method of Separation of Variables

In an attempt to solve a large class of mixed-type problems, we now explore a valuable mathematical technique called *the method of separation of variables*.   Detailed examples of this method will be presented after the following general description.

Consider the problem of finding a solution $u = f(x,y)$ on $\bar{R}$ of the mixed-type problem

$$u_{xx} - u_{yy} = 0, \qquad\qquad \text{on } R, \tag{4.44}$$

$$f(x,0) = g(x), \qquad\qquad 0 \le x \le \pi, \tag{4.45}$$

$$\frac{\partial f(x,0)}{\partial y} = 0, \qquad\qquad 0 \le x \le \pi, \tag{4.46}$$

$$f(0,y) = f(\pi,y) = 0, \qquad y \ge 0. \tag{4.47}$$

If $g(x) \equiv 0$ for $0 \le x \le \pi$, then $f(x,y) \equiv 0$ on $\bar{R}$ is such a solution. Assume then $g(x) \not\equiv 0$ on $0 \le x \le \pi$ and consider a *nontrivial* solution of the form

$$f(x,y) = X(x)Y(y), \tag{4.48}$$

where X is a function of x alone and Y is a function of y alone. Substitution of (4.48) into p.d.e. (4.44) yields

$$X(x)\frac{d^2Y}{dy^2} - Y(y)\frac{d^2X}{dx^2} \equiv 0, \qquad (x,y) \in R. \tag{4.49}$$

For values of x, y for which $X(x) \neq 0$, $Y(y) \neq 0$, (4.49) implies

$$\left[\frac{d^2X(x)}{dx^2} \middle/ X(x)\right] \equiv \left[\frac{d^2Y(y)}{dy^2} \middle/ Y(y)\right]. \tag{4.50}$$

Since the left side of (4.50) is a function of x alone, while the right-hand side is a function of y alone, it follows necessarily that, for λ constant,

$$\frac{d^2X(x)}{dx^2} = \lambda X(x), \tag{4.51}$$

$$\frac{d^2Y(y)}{dy^2} = \lambda Y(y). \tag{4.52}$$

By virtue of the particular forms of (4.48), (4.51), and (4.52), one says that the variables x, y of (4.44) have been separated and consideration of the pair of ordinary differential equations (4.51) and (4.52) will replace consideration of the partial differential equation (4.44).

Since (4.48) is assumed to be nontrivial, (4.47) implies

$$X(0) = X(\pi) = 0. \tag{4.53}$$

However, ordinary differential equation (4.51) can now be solved subject to (4.53) by consideration of the three cases $\lambda = 0$, $\lambda > 0$, $\lambda < 0$.

CASE 1. If $\lambda = 0$, the general solution of (4.51) is $X = c_1 + c_2 x$. But (4.53) then implies $c_1 = c_2 = 0$, so that $X(x) \equiv 0$, which contradicts the assumption that (4.48) is nontrivial.

CASE 2. If $\lambda > 0$, the general solution of (4.51) is $X = c_1 e^{\sqrt{\lambda}x} + c_2 e^{-\sqrt{\lambda}x}$. But (4.53) again implies $c_1 = c_2 = 0$, so that $X(x) \equiv 0$, which contradicts the assumption that (4.48) is nontrivial.

CASE 3. If $\lambda < 0$, the general solution of (4.51) is

$$X(x) = c_1 \cos [(-\lambda)^{\frac{1}{2}}x] + c_2 \sin [(-\lambda)^{\frac{1}{2}}x],$$

and (4.53) then implies, provided $\lambda = -k^2$, $k = 1, 2, \ldots$, that $c_1 = 0$ and that c_2 is arbitrary. Without loss of generality, let $c_2 = 1$.

Cases 1, 2, 3, above, indicate, then, that in order to arrive at a nontrivial solution of type (4.48) of p.d.e. (4.44), it is necessary that λ have the values

$$\lambda_k = -k^2, \qquad k = 1, 2, 3, \ldots, \tag{4.54}$$

and that corresponding to each such λ is the function

$$X_k(x) = \sin kx; \qquad k = 1, 2, \ldots, \tag{4.55}$$

which is a solution of (4.51) subject to (4.53).

The values λ_k defined by (4.54) are called eigenvalues, or characteristic values, and the functions X_k defined by (4.55) are called eigenfunctions, or characteristic functions.

Since the values of λ have been restricted to those given in (4.54), one must examine the ramifications on ordinary differential equation (4.52). This equation now becomes

$$\frac{d^2 Y(y)}{dy^2} + k^2 Y(y) = 0, \qquad k = 1, 2, \ldots, \tag{4.56}$$

and has, for each k, a general solution of the form

$$Y_k(y) = A_k \cos ky + B_k \sin ky, \tag{4.57}$$

where A_k, B_k, $k = 1, 2, \ldots$, are arbitrary constants. Thus (4.55) and (4.57) imply that each of the functions

$$u_k(x,y) = X_k(x)Y_k(y) \equiv \sin kx(A_k \cos ky + B_k \sin ky), \quad k = 1, 2, 3, \ldots, \tag{4.58}$$

is a solution of (4.44) which satisfies (4.47).

Consider then the series

$$u = f(x,y) \equiv \sum_{k=1}^{\infty} \sin kx(A_k \cos ky + B_k \sin ky) \tag{4.59}$$

and let us determine the constants A_k, B_k, $k = 1, 2, \ldots$, so that (4.59) satisfies (4.45) and (4.46). Assuming that (4.59) converges, has continuous first derivative with respect to y, and that the series which results from such differentiation also converges, then (4.45), (4.46), and (4.59) imply

$$f(x,0) = \sum_{k=1}^{\infty} A_k \sin kx \equiv g(x) \tag{4.60}$$

$$\frac{\partial f(x,0)}{\partial y} = \sum_{k=1}^{\infty} kB_k \sin kx \equiv 0. \tag{4.61}$$

Setting $B_k = 0$; $k = 1, 2, \ldots$, will satisfy (4.61), while from (4.60) it follows that the A_k's must be the *coefficients of the Fourier half-range sine series for $g(x)$*. Thus, A_k, B_k, $k = 1, 2, 3, \ldots$ are determined and (4.59) presents itself as a possible solution of the mixed-type problem in the form

$$f(x,y) = \sum_{k=1}^{\infty} A_k \sin kx \cos ky. \tag{4.62}$$

The details involved in verifying whether or not (4.62) converges, is termwise differentiable, and is actually a solution of the given problem will now be illustrated by three examples.

Example 1. Find a solution $u = f(x,y)$ on $\bar{R}$ of the mixed-type problem

$$u_{xx} - u_{yy} = 0, \qquad\qquad \text{on } R, \qquad\qquad (4.63)$$

$$f(x,0) = \sin x, \qquad\qquad 0 \le x \le \pi, \qquad\qquad (4.64)$$

$$\frac{\partial f(x,0)}{\partial y} = 0, \qquad\qquad 0 \le x \le \pi, \qquad\qquad (4.65)$$

$$f(0,y) = f(\pi,y) = 0, \qquad y \ge 0. \qquad\qquad (4.66)$$

Solution. The Fourier half-range sine series for $\sin x$ is simply $\sin x$. Hence, $A_1 = 1$, $A_k = 0$; $k = 2, 3, \ldots$, so that (4.62) becomes

$$f(x,y) = \sin x \cos y, \qquad\qquad (4.67)$$

which is easily verified to be the desired solution. Of course, since $f(x,y) = \sin x \cos y \in C^2$ on $\bar{R}$, it is the unique such solution of class C^2.

Example 2. Find a solution $u = f(x,y)$ on $\bar{R}$ of the mixed-type problem

$$u_{xx} - u_{yy} = 0, \qquad\qquad \text{on } R, \qquad\qquad (4.68)$$

$$f(x,0) = 1 - \cos 2x, \qquad 0 \le x \le \pi, \qquad\qquad (4.69)$$

$$\frac{\partial f(x,0)}{\partial y} = 0, \qquad\qquad 0 \le x \le \pi, \qquad\qquad (4.70)$$

$$f(0,y) = f(\pi,y) = 0, \qquad y \ge 0. \qquad\qquad (4.71)$$

Solution. The desired Fourier half-range sine series for $g(x) = 1 - \cos 2x$ is given by $\sum_1^\infty b_k^* \sin kx$, where

$$\left. \begin{aligned} &b_1^* = \frac{16}{3\pi}, \qquad b_2^* = 0, \\[2mm] &b_k^* = \frac{2}{\pi}\left[\frac{1}{k} - \frac{\cos k\pi}{k} + \frac{\cos(k-2)\pi}{2(k-2)} + \frac{\cos(k+2)\pi}{2(k+2)} \right. \\[2mm] &\qquad\qquad \left. - \frac{1}{2(k-2)} - \frac{1}{2(k+2)} \right], \qquad k = 3, 4, \ldots \end{aligned} \right\} \qquad (4.72)$$

Thus (4.62) becomes

$$f(x,y) = \sum_{k=1}^\infty b_k^* \sin kx \cos ky, \qquad\qquad (4.73)$$

where b_k^*, $k = 1, 2, 3, \ldots$, are given by (4.72).

Since

$$\sin kx \cos ky \equiv \tfrac{1}{2}[\sin(kx + ky) + \sin(kx - ky)],$$

if it can be shown that

$$f(x,y) = \frac{1}{2} \sum_{k=1}^{\infty} b_k^* \sin(kx + ky) + \frac{1}{2} \sum_{k=1}^{\infty} b_k^* \sin(kx - ky) \quad (4.74)$$

is a solution of the given problem, then (4.73) will also be a solution of the given problem. This will now be done.

By (4.72), it follows that (4.74) can be written

$$f(x,y) = \tfrac{1}{2}[1 - \cos 2(x + y)] + \tfrac{1}{2}[1 - \cos 2(x - y)], \quad (4.74')$$

which is of the form

$$f_1(x + y) + f_2(x - y),$$

so that by Theorem 2.5 and Theorem 4.1, $f(x,y)$ satisfies the homogeneous wave equation on R.

Verification that (4.73) satisfies conditions (4.69) to (4.71) follows easily. Finally, since (4.74') is of class C^2 on $\bar{R}$, it is, by Theorem 4.5, the unique solution of class C^2 on $\bar{R}$ of the given mixed-type problem.

Note that Example 2, above, provides a technique by which one can convert a Fourier series type solution of a mixed-type problem to a solution of type (4.4).

Example 3. Find a solution $u = f(x,y)$ on $\bar{R}$ of the mixed-type problem

$$u_{xx} - u_{yy} = 0, \qquad\qquad \text{on } R, \qquad\qquad (4.75)$$

$$f(x,0) = \begin{cases} 0, & 0 \le x < \dfrac{\pi}{4} \\[2mm] 2, & \dfrac{\pi}{4} < x < \dfrac{3\pi}{4} \\[2mm] 0, & \dfrac{3\pi}{4} < x < \pi \\[2mm] 1, & x = \dfrac{\pi}{4}, \dfrac{3\pi}{4}, \end{cases} \qquad (4.76)$$

$$\frac{\partial f(x,0)}{\partial y} = 0, \qquad\qquad 0 \le x \le \pi, \qquad (4.77)$$

$$f(0,y) = f(\pi,y) = 0, \qquad y \ge 0. \qquad (4.78)$$

Discussion. The half-range sine series for (4.76) is $\sum_{k=1}^{\infty} b_k^* \sin kx$, where

$$b_k^* = \frac{4}{\pi k}\left[\cos\frac{k\pi}{4} - \cos\frac{3\pi k}{4}\right], \qquad k = 1, 2, \ldots. \qquad (4.79)$$

Thus, (4.73) takes the form

$$f(x,y) = \sum_{k=1}^{\infty} b_k^* \sin kx \cos ky, \tag{4.80}$$

where b_k^*, $k = 1, 2, \ldots$, are defined by (4.79). However, it does not necessarily follow as in the fashion described in Example 2, above, that (4.80) is a solution of the given mixed-type problem. The reason for this is that (4.76) does not satisfy the sufficiency conditions set forth for differentiability in Theorem 2.5. A more detailed examination, then, of this and similar problems must be reserved for an advanced treatment (e.g., Sobolew [44]), and the reader is cautioned to be fastidious in applying and interpreting the method of separation of variables.

EXERCISES

4.1. For Cauchy Problem I, find the domain of dependence with respect to each of the following points: $(0,1)$, $(2,1)$, $(1,2)$, $(3,4)$, $(-3,4)$, $(-4,3)$.
  (*Ans.* $-1 \leq x \leq 1$, $1 \leq x \leq 3$, $-1 \leq x \leq 3$, $-1 \leq x \leq 7$, $-7 \leq x \leq 1$,
                                                                            $-7 \leq x \leq -1$)

4.2. Find the solution on E^2 of each of the following Cauchy problems (type I) and check each answer:

(a) $u_{xx} - u_{yy} = 0$, $f(x,0) = 1$, $f_y(x,0) = 0$
(b) $u_{xx} - u_{yy} = 0$, $f(x,0) = 0$, $f_y(x,0) = 1$
(c) $u_{xx} - u_{yy} = 0$, $f(x,0) = x^2$, $f_y(x,0) = 1$
(d) $u_{xx} - u_{yy} = 0$, $f(x,0) = x$, $f_y(x,0) = x$
(e) $u_{xx} - u_{yy} = 0$, $f(x,0) = \sin x$, $f_y(x,0) = \cos x$
(f) $u_{xx} - u_{yy} = 0$, $f(x,0) = e^x$, $f_y(x,0) = (1 + x^2)^{-1}$
(g) $u_{xx} - u_{yy} = 0$, $f(x,0) = \log(1 + x^2)$, $f_y(x,0) = 0$
(h) $u_{xx} - u_{yy} = 0$, $f(x,0) = 1 + x + x^2$, $f_y(x,0) = \cosh x$.

4.3. Find the solution on E^2 of each of the following Cauchy problems (type II) and check each answer:

(a) $u_{xx} - u_{yy} = 1$, $f(x,0) = 1$, $f_y(x,0) = 0$
(b) $u_{xx} - u_{yy} = -2x$, $f(x,0) = 0$, $f_y(x,0) = 1$
(c) $u_{xx} - u_{yy} = 4y$, $f(x,0) = x^2$, $f_y(x,0) = 1$
(d) $u_{xx} - u_{yy} = x + 2y$, $f(x,0) = x$, $f_y(x,0) = x$
(e) $u_{xx} - u_{yy} = xy$, $f(x,0) = \cos x$, $f_y(x,0) = \sin x$
(f) $u_{xx} - u_{yy} = 3x^2y - y^2x$, $f(x,0) = e^{-x}$, $f_y(x,0) = -(1 + x^2)^{-1}$
(g) $u_{xx} - u_{yy} = e^x + \sin y$, $f(x,0) = \log(1 + 2x^2)$, $f_y(x,0) = 0$
(h) $u_{xx} - u_{yy} = e^{x+y} + \cos xy$, $f(x,0) = 1 - x^2$, $f_y(x,0) = \sinh x$.

4.4. Denote by $\bar{R}$ the set of points (x,y) whose coordinates satisfy $0 \leq x \leq \pi$, $y \geq 0$, and by R the set of points (x,y) whose coordinates satisfy $0 < x < \pi$,

$y > 0$.   Then find, if possible, a solution on $\bar{R}$ for each of the following mixed-type problems:

(a) $u_{xx} - u_{yy} = 0$,   on R,
$\quad$ $f(x,0) = \sin^2 x$, $0 \leq x \leq \pi$,
$\quad$ $f_y(x,0) = 0$, $0 \leq x \leq \pi$,
$\quad$ $f(0,y) = f(\pi,y) = 0$, $y \geq 0$.

(b) $u_{xx} - u_{yy} = 0$,   on R,
$\quad$ $f(x,0) = \sin^3 x$, $0 \leq x \leq \pi$,
$\quad$ $f_y(x,0) = 0$, $0 \leq x \leq \pi$,
$\quad$ $f(0,y) = f(\pi,y) = 0$, $y \geq 0$.

(c) $u_{xx} - u_{yy} = 0$,   on R,
$\quad$ $f(x,0) = -\pi^2 x^2 + 2\pi x^3 - x^4$, $0 \leq x \leq \pi$,
$\quad$ $f_y(x,0) = 0$, $0 \leq x \leq \pi$,
$\quad$ $f(0,y) = f(\pi,y) = 0$, $y \geq 0$.

(d) $u_{xx} - u_{yy} = 0$,   on R,
$\quad$ $f(x,0) = 1 - \cos 4x$, $0 \leq x \leq \pi$,
$\quad$ $f_y(x,0) = 0$, $0 \leq x \leq \pi$,
$\quad$ $f(0,y) = f(\pi,y) = 0$, $y \geq 0$.

(e) $u_{xx} - u_{yy} = 0$,   on R,
$\quad$ $f(x,0) = \begin{cases} x, \\ 0, \end{cases}$ $\begin{array}{l} 0 \leq x < \pi \\ x = \pi, \end{array}$
$\quad$ $f_y(x,0) = 0$, $0 \leq x \leq \pi$
$\quad$ $f(0,y) = f(\pi,y) = 0$, $y \geq 0$.

(f) $u_{xx} - u_{yy} = 0$,   on R,

$$f(x,0) = \begin{cases} 1, & 0 < x < \dfrac{\pi}{2} \\[2mm] -1, & \dfrac{\pi}{2} < x < \pi \\[2mm] 0, & x = 0, \dfrac{\pi}{2}, \pi, \end{cases}$$

$\quad$ $f_y(x,0) = 0$, $0 \leq x \leq \pi$,
$\quad$ $f(0,y) = f(\pi,y) = 0$, $y \geq 0$.

4.5. Find, if possible, a solution on $\bar{R}$ for each of the following mixed-type problems:

(a) $u_{xx} - u_{yy} = 0$,   on R,
$\quad$ $f(x,0) = 2 \sin x \cos x$, $0 \leq x \leq 1$,
$\quad$ $f_y(x,0) = 0$, $0 \leq x \leq 1$,
$\quad$ $f(0,y) = f(1,y) = 0$, $y \geq 0$.

(b) $u_{xx} - u_{yy} = 0$,   on R,
$\quad$ $f(x,0) = x \sin x$, $0 \leq x \leq 2$,
$\quad$ $f_y(x,0) = 0$, $0 \leq x \leq 2$,
$\quad$ $f(0,y) = f(2,y) = 0$, $y \geq 0$.

(c) $u_{xx} - u_{yy} = 0,$ $\qquad\qquad$ on $R,$

$\qquad f(x,0) = \sin 4x,$ $\qquad\quad 0 \leq x \leq 4\pi,$

$\qquad f_y(x,0) = 0,$ $\qquad\qquad 0 \leq x \leq 4\pi,$

$\qquad f(0,y) = f(4\pi,y) = 0,$ $\qquad y \geq 0.$

4.6. Let $(x,y) \in \bar{R}$ if and only if $0 \leq x \leq \pi, y \geq 0$ and let $(x,y) \in R$ if and only if $0 < x < \pi, y > 0.$

(a) On $\bar{R}$ find a function $u = f(x,y)$ which satisfies

$\qquad\qquad u_{xx} - u_{yy} = 0,$ $\qquad\qquad$ on $R,$

$\qquad\qquad\quad f(x,0) = \sin 2x,$ $\qquad 0 \leq x \leq \pi,$

$\qquad\qquad\quad f_y(x,0) = 0,$ $\qquad\qquad 0 \leq x \leq \pi,$

$\qquad\qquad\quad f(0,y) = f(\pi,y) = 0,$ $\qquad y \geq 0.$

(b) On $\bar{R}$ find a function $u = f(x,y)$ which satisfies

$\qquad\qquad u_{xx} - u_{yy} = 0,$ $\qquad\qquad$ on $R,$

$\qquad\qquad\quad f(x,0) = 0,$ $\qquad\qquad 0 \leq x \leq \pi,$

$\qquad\qquad\quad f_y(x,0) = \sin x,$ $\qquad 0 \leq x \leq \pi,$

$\qquad\qquad\quad f(0,y) = f(\pi,y) = 0,$ $\qquad y \geq 0.$

(c) Using the results of (a) and (b), above, find a function $u = f(x,y)$ which satisfies

$\qquad\qquad u_{xx} - u_{yy} = 0,$ $\qquad\qquad$ on $R,$

$\qquad\qquad\quad f(x,0) = \sin 2x,$ $\qquad 0 \leq x \leq \pi,$

$\qquad\qquad\quad f_y(x,0) = \sin x,$ $\qquad 0 \leq x \leq \pi,$

$\qquad\qquad\quad f(0,y) = f(\pi,y) = 0,$ $\qquad y \geq 0.$

4.7. Let $(x,y) \in R$ if and only if $0 < x < \pi, y > 0.$ Then extend the discussion of the method of separation of variables to the mixed-type problem

$\qquad\qquad u_{xx} - u_{yy} = 0,$ $\qquad\qquad$ on $R,$

$\qquad\qquad\quad f(x,0) = g_1(x),$ $\qquad 0 \leq x \leq \pi,$

$\qquad\qquad\quad f_y(x,0) = g_2(x),$ $\qquad 0 \leq x \leq \pi,$

$\qquad\qquad\quad f(0,y) = f(\pi,y) = 0,$ $\qquad y \geq 0.$

4.8. Denote by $\bar{R}$ the set of points (x,y) whose coordinates satisfy $0 \leq x \leq \pi,$ $y \geq 0$ and by R the set of points (x,y) whose coordinates satisfy $0 < x < \pi,$ $y > 0.$ Then, by means of the method of separation of variables, find, if possible, a solution $u = f(x,y)$ on $\bar{R}$ of each of the following modified mixed-type problems:

(a) $u_{xx} - 4u_{yy} = 0,$ $\qquad\qquad$ on $R,$

$\qquad f(x,0) = \sin x,$ $\qquad\quad 0 \leq x \leq \pi,$

$\qquad f_y(x,0) = 0,$ $\qquad\qquad 0 \leq x \leq \pi,$

$\qquad f(0,y) = f(\pi,y) = 0,$ $\qquad y \geq 0.$

(b) $u_{xx} - c^2 u_{yy} = 0, (c \neq 0),$ $\qquad$ on $R,$

$\qquad\quad f(x,0) = 1 - \cos 2x,$ $\qquad 0 \leq x \leq \pi,$

$\qquad\quad f_y(x,0) = 0,$ $\qquad\qquad 0 \leq x \leq \pi,$

$\qquad\quad f(0,y) = f(\pi,y) = 0,$ $\qquad y \geq 0.$

4.9.   For $a > 0$, $b > 0$, let $(x,y) \in \bar{R}$ if and only if $0 \le x \le a$, $0 \le y \le b$, and let $(x,y) \in R$ if and only if $0 < x < a$, $0 < y < b$.   Consider the problem of finding a function $u = f(x,y)$ on $\bar{R}$ which is a solution on R of the partial differential equation

$$\frac{\partial^2 u}{\partial x \, \partial y} = G(x,y,u)$$

and which satisfies the conditions

$$f(x,0) = 0, \qquad 0 \le x \le a,$$
$$f(0,y) = 0, \qquad 0 \le y \le b.$$

Assuming that $f(x,y)$ exists and is unique, set

$$f^{(n+1)}(x,y) = \int_0^x \int_0^y G(\xi,\eta,f^{(n)}(\xi,\eta)) \, d\xi \, d\eta, \qquad n = 0, 1, 2, \ldots,$$

where $f^{(0)}(x,y)$ is an initial approximation of $f(x,y)$ and find sufficient conditions so that on $\bar{R}$

$$\lim_{n \to \infty} f^{(n)}(x,y) = f(x,y).$$

Chapter 5

THE POTENTIAL EQUATION

The physical study of potential leads quite naturally to the consideration of an elliptic partial differential equation called the *potential*, or *Laplace's*, equation.

Definition 5.1. The equation

$$u_{xx} + u_{yy} = 0, \tag{5.1}$$

written alternately as

$$\Delta u \equiv \frac{\partial^2 u}{\partial x^2} + \frac{\partial^2 u}{\partial y^2} = 0, \tag{5.1'}$$

is called the potential equation.

Definition 5.2. If N is a given plane point set, then a function $u = f(x,y)$ which is of class C^2 and is a solution of p.d.e. (5.1) on N is said to be harmonic on N.

Examples. On E^2, $u = 4$, $u = x$, $u = y$, $u = x^2 - y^2$, $u = \frac{1}{2}xy^2 - \frac{1}{6}x^3$, $u = \frac{1}{6}(xy^3 - x^3y)$ are harmonic.

We first state and prove a rather obvious theorem which was intrinsic to, but not explicitly stated in, the discussion of Chap. 3. It is stated here only for reference purposes throughout this chapter.

Theorem 5.1. Let $u = f(x,y)$ be harmonic on set M. For α, β constants, let the translation $x = \bar{x} + \alpha$, $y = \bar{y} + \beta$ map M into $\bar{M}$ and let $f(x,y)$ transform into $\bar{f}(\bar{x},\bar{y})$. Then $\bar{f}(\bar{x},\bar{y})$ is harmonic on $\bar{M}$.

Proof.

$$\frac{\partial^2 f}{\partial \bar{x}^2} + \frac{\partial^2 f}{\partial \bar{y}^2} = \frac{\partial}{\partial \bar{x}}\left(\frac{\partial f}{\partial x}\frac{\partial x}{\partial \bar{x}} + \frac{\partial f}{\partial y}\frac{\partial y}{\partial \bar{x}}\right) + \frac{\partial}{\partial \bar{y}}\left(\frac{\partial f}{\partial x}\frac{\partial x}{\partial \bar{y}} + \frac{\partial f}{\partial y}\frac{\partial y}{\partial \bar{y}}\right)$$

$$= \frac{\partial}{\partial \bar{x}}\left(\frac{\partial f}{\partial x}\right) + \frac{\partial}{\partial \bar{y}}\left(\frac{\partial f}{\partial y}\right)$$

$$= \frac{\partial^2 f}{\partial x^2} + \frac{\partial^2 f}{\partial y^2}$$

$$= 0,$$

and the theorem is proved.

As with the homogeneous wave equation, the examples provided with Definition 5.2 indicate that further restrictions must be imposed on a solution of the potential equation if that solution is to be unique. However, since p.d.e. (5.1) has no real characteristics, one might surmise that, in general, there does not exist a real-valued function $u = f(x,y)$ which satisfies the potential equation on E^2 and which satisfies Cauchy conditions (4.15) and (4.16). That this is actually the case can be demonstrated with the aid of the theory of complex-valued functions (see, for example, John [25]). Motivated then by physical considerations, we shall consider with respect to the potential equation a class of problems called Dirichlet, or boundary-value, problems.

5.1 The Dirichlet Problem

Definition 5.3. In E^2, let R be a simply connected, bounded region whose boundary B is a contour. Set $\bar{R} = R \cup B$. If $g(x,y)$ is a given function which is defined and continuous on B, then the Dirichlet problem is that of determining a function $u = f(x,y)$ which is:

(a) defined and continuous on $\bar{R}$,
(b) harmonic on R, and
(c) identical with $g(x,y)$ on B.

Example. The problem of determining a function $u = f(x,y)$ such that:

(a) $f(x,y)$ is continuous at all points (x,y) whose coordinates satisfy $x^2 + y^2 \leq 1$,

(b) $f(x,y)$ is harmonic at all points (x,y) whose coordinates satisfy $x^2 + y^2 < 1$,

(c) $f(x,y)$ coincides with $g(x,y) = 1 + x + y$ at all points (x,y) whose
   coordinates satisfy $x^2 + y^2 = 1$,

is a Dirichlet problem.

Note that if in Definition 5.3, B is given parametrically by
$x = h_1(t)$, $y = h_2(t)$, then $g(x,y)$ may be given in terms of parameter
t by $g(h_1(t),h_2(t)) = H(t)$.   In the above example, if the curve with
equation $x^2 + y^2 = 1$ is parametrized by $x = \cos t$, $y = \sin t$, $-\pi \le$
$t \le \pi$, one may set $g(x,y) = 1 + x + y = 1 + \cos t + \sin t = H(t)$.

Relative to the Dirichlet problem, consider first, in a heuristic
fashion, a function $u = f(x,y)$ which is continuous on $\bar{R}$ and
harmonic on R.   Let point P, with coordinates (x_1,y_1), be in R, and
suppose $u_{xx} \ne 0$ at P.   Consider, say, $u_{xx} > 0$ at P.   Then the
plane curve, determined by the surface whose equation is $u = f(x,y)$
and the plane whose equation is $y = y_1$, is concave up at P.   Thus,
$u = f(x,y)$ *cannot* assume a maximum value at (x_1,y_1).   However,
$u_{xx} > 0$ at P implies, by (5.1), that $u_{yy} < 0$ at P, and one can readily
conclude that $u = f(x,y)$ *cannot* assume a minimum value at (x_1,y_1).
Hence if $u_{xx} \ne 0$ at *any* point of R, then $u = f(x,y)$ *must assume both
its maximum and its minimum values on the boundary B of R.*   That
this latter conclusion, derived from very special considerations, is
actually valid *in general* is the essence of the following two theorems.

Theorem 5.2.   Maximum Principle.   Let R be a simply connected, bounded region whose boundary B is a contour.   Set $\bar{R} =$
$R \cup B$.   If $u = f(x,y)$ is harmonic on R and continuous on $\bar{R}$, then
$f(x,y)$ attains its maximum value on B.

Proof.   Since each of the point sets B and $\bar{R}$ is a closed, bounded
set in E^2, it follows that since $f(x,y)$ is continuous on B and on $\bar{R}$,
the function assumes on B a maximum m and on $\bar{R}$ a maximum M.

Assume that $M > m$.   Let D be the diameter of $\bar{R}$ and let (x_1,y_1)
be a point of R such that $f(x_1,y_1) = M$.   Consider

$$w(x,y) = f(x,y) + \frac{M - m}{2D^2} [(x - x_1)^2 + (y - y_1)^2]. \qquad (5.2)$$

Note that for $(x,y) \in R$, $(x - x_1)^2 + (y - y_1)^2 < D^2$ and that
$w(x_1,y_1) = f(x_1,y_1) = M$.   Note also that if $(x,y) \in B$, then

$$w(x,y) \le m + \frac{M - m}{2} = \tfrac{1}{2}(m + M) < M.$$

Hence, $w(x,y)$, like $f(x,y)$, must attain its maximum at a point of R.
Let $(x_2,y_2) \in R$ be a point at which $w(x,y)$ attains its maximum.

Thus

$$\frac{\partial^2 w(x_2, y_2)}{\partial x^2} \leq 0, \qquad \frac{\partial^2 w(x_2, y_2)}{\partial y^2} \leq 0, \qquad (5.3)$$

so that

$$\frac{\partial^2 w(x_2, y_2)}{\partial x^2} + \frac{\partial^2 w(x_2, y_2)}{\partial y^2} \leq 0. \qquad (5.4)$$

However, from (5.2), one has

$$\frac{\partial^2 w(x_2, y_2)}{\partial x^2} + \frac{\partial^2 w(x_2, y_2)}{\partial y^2} =$$

$$\frac{\partial^2 f(x_2, y_2)}{\partial x^2} + \frac{\partial^2 f(x_2, y_2)}{\partial y^2} + \frac{M - m}{D^2} = \frac{M - m}{D^2} > 0, \qquad (5.5)$$

which contradicts (5.4). Hence, the assumption $M > m$ is incorrect. However, since M is certainly not smaller than m, it follows that $M = m$ and the theorem is proved.

Theorem 5.3. Minimum Principle. Let R be a simply connected, bounded region whose boundary B is a contour. Set $\bar{R} = R \cup B$. If $u = f(x,y)$ is harmonic on R and continuous on R, then $f(x,y)$ attains its minimum value on B.

Proof. The theorem follows directly by applying Theorem 5.2 to the harmonic function $-f(x,y)$.

Example. Let R be the set of all points (x,y) whose coordinates satisfy $x^2 + y^2 < 1$. Then B is the set of all points (x,y) whose coordinates satisfy $x^2 + y^2 = 1$. The function $f(x,y) = x^2 - y^2$ is harmonic on R and continuous on $\bar{R} = R \cup B$. Then $f(x,y)$ must attain its maximum and its minimum values on B, and, as is readily verified, $f(x,y)$ attains its maximum value of 1 at the point $(1,0) \in B$ and its minimum value of -1 at the point $(0,1) \in B$. These points, of course, are not necessarily unique.

Theorem 5.4. Let R be a simply connected, bounded region whose boundary B is a contour. Let $f_1(x,y)$, $f_2(x,y)$ be harmonic on R and continuous on $\bar{R} = R \cup B$. If $f_1(x,y) \equiv f_2(x,y)$ on B, then $f_1(x,y) \equiv f_2(x,y)$ on $\bar{R}$.

Proof. The function

$$f(x,y) = f_1(x,y) - f_2(x,y) \qquad (5.6)$$

is harmonic on R and continuous on $\bar{R}$. Hence, $f(x,y)$ takes on its maximum and its minimum values on B. But $f(x,y) \equiv 0$ on B, so that $f(x,y) \equiv 0$ on $\bar{R}$. Hence, $f_1(x,y) \equiv f_2(x,y)$ on $\bar{R}$ and the theorem is proved.

Theorem 5.5. If the Dirichlet problem has a solution, then that solution is unique.

Proof. The proof is an immediate consequence of Theorem 5.4.

Theorem 5.6. If the Dirichlet problem has a solution, then that solution depends continuously on the boundary function $g(x,y)$.

Proof. Let $f_1(x,y)$, $f_2(x,y)$ be solutions of Dirichlet problems on $\bar{R} = R \cup B$ with boundary functions $g_1(x,y)$, $g_2(x,y)$, respectively. Suppose that on B,

$$|g_1(x,y) - g_2(x,y)| < \epsilon,$$

or, equivalently, that on B,

$$-\epsilon < g_1(x,y) - g_2(x,y) < \epsilon.$$

Then the function $f(x,y)$ defined on $\bar{R}$ by

$$f(x,y) = f_1(x,y) - f_2(x,y)$$

is a solution on $\bar{R}$ of the Dirichlet problem with boundary function $g_1(x,y) - g_2(x,y)$. Hence, $f(x,y)$ assumes its maximum and its minimum values on B, so that

$$-\epsilon < f_1(x,y) - f_2(x,y) < \epsilon$$

on B, and hence on $\bar{R}$, so that the theorem is proved.

Theorem 5.7. Let R be a simply connected, bounded region whose boundary B is a contour. Let $\{f_n(x,y)\}$ be a sequence of functions each of which is continuous on $\bar{R} = R \cup B$ and harmonic on R. If $\{f_n(x,y)\}$ converges uniformly on B, then $\{f_n(x,y)\}$ converges uniformly on $\bar{R}$.

Proof. By assumption, $\{f_n(x,y)\}$ converges uniformly on B. Thus, given $\epsilon > 0$, there exists N such that $m, n > N$ imply $|f_n(x,y) - f_m(x,y)| < \epsilon$ on B. By Theorems 5.2 and 5.3, it follows then that for all $m, n > N$, $|f_n(x,y) - f_m(x,y)| < \epsilon$ on $\bar{R}$, from which the theorem results.

5.2 Poisson's Integral Formula

We now set upon the arduous task of establishing the *existence* of the solution $u = f(x,y)$ of the Dirichlet problem by first considering R to be the interior of the unit circle. In this elementary case the solution can be given explicitly in terms of the Poisson integral.

Theorem 5.8. For the Dirichlet problem, let R be the interior of the unit circle, and let its boundary B be given parametrically by $x = \cos \phi, y = \sin \phi, -\pi \leq \phi \leq \pi$. Let $g(x,y) = g(\cos \phi, \sin \phi) = H(\phi)$, where $H(\phi)$ is continuous on $-\pi \leq \phi \leq \pi$ and satisfies $H(-\pi) = H(\pi)$. If $u = f(x,y)$ is defined on $\bar{R} = R \cup B$ by

$$f(x,y) = g(x,y) \equiv H(\phi), \qquad x^2 + y^2 = 1, \qquad (5.7)$$

$$f(x,y) = \frac{1}{2\pi} \int_{-\pi}^{\pi} \frac{H(\xi)(1 - x^2 - y^2)}{1 - 2x \cos \xi - 2y \sin \xi + x^2 + y^2} \, d\xi, \qquad x^2 + y^2 < 1,$$
$$(5.8)$$

then $u = f(x,y)$ is the unique solution of the Dirichlet problem under consideration.

Proof. By Theorem 5.5, one need only show that $f(x,y)$ is *a* solution of the problem and the theorem will follow. For this purpose it will be convenient to consider (5.8) in the alternate form

$$f(x,y) = \frac{1}{2\pi} \int_{-\pi}^{\pi} \frac{H(\xi)(1 - r^2)}{1 - 2r \cos (\phi - \xi) + r^2} \, d\xi, \qquad x^2 + y^2 < 1, \quad (5.8')$$

where (r,ϕ) are polar coordinates of (x,y). Equivalent integrals (5.8) and (5.8') will both be called the Poisson integral.

For $r = \sqrt{x^2 + y^2} < 1$, one has

$$1 - 2x \cos \xi - 2y \sin \xi + x^2 + y^2 = 1 - 2r \cos (\phi - \xi) + r^2$$
$$\geq 1 - 2r + r^2$$
$$= (1 - r)^2$$
$$> 0,$$

so that the denominator of the integrand of Poisson integral (5.8) is never zero. It follows readily then that Poisson integral (5.8) can be differentiated freely with respect to x and y and is harmonic on R. Moreover, since for each r_0 which satisfies $0 \leq r_0 < 1$, (5.8) implies that f_x, f_y are continuous for all (x,y) whose coordinates satisfy $x^2 + y^2 \leq r_0^2$, it follows that Poisson integral (5.8) is continuous on R.

In order to show that $f(x,y)$, defined by (5.7) and (5.8), is the solution of the Dirichlet problem under consideration, it is only necessary to show that $f(x,y)$ is continuous on $\bar{R}$. This will be done by showing that, uniformly in ϕ,

$$\lim_{r \to 1^-} \frac{1}{2\pi} \int_{-\pi}^{\pi} \frac{H(\xi)(1 - r^2)}{1 - 2r \cos (\phi - \xi) + r^2} \, d\xi = H(\phi). \qquad (5.9)$$

Assume first that $H(\phi) \in C^2$ *on* $-\pi \leq \phi \leq \pi$. Then $H(\phi)$ has a Fourier series representation

$$H(\phi) = \frac{a_0}{2} + \sum_{n=1}^{\infty} (a_n \cos n\phi + b_n \sin n\phi), \qquad (5.10)$$

where

$$a_n = \frac{1}{\pi} \int_{-\pi}^{\pi} H(\xi) \cos n\xi \, d\xi, \qquad n = 0, 1, 2, \ldots, \tag{5.11}$$

$$b_n = \frac{1}{\pi} \int_{-\pi}^{\pi} H(\xi) \sin n\xi \, d\xi, \qquad n = 1, 2, \ldots . \tag{5.12}$$

By Corollary 2, Theorem 2.6, the series (5.10) converges absolutely and uniformly. Hence, for $0 \le r < 1$, the series

$$H^{(r)}(\phi) = \frac{a_0}{2} + \sum_{n=1}^{\infty} [r^n(a_n \cos n\phi + b_n \sin n\phi)] \tag{5.13}$$

converges absolutely and uniformly. By (5.11) and (5.12), one has

$$H^{(r)}(\phi) = \frac{1}{2\pi} \int_{-\pi}^{\pi} H(\xi) \, d\xi + \frac{1}{\pi} \sum_{n=1}^{\infty} \left[r^n \left(\cos n\phi \int_{-\pi}^{\pi} H(\xi) \cos n\xi \, d\xi \right.\right.$$
$$\left.\left. + \sin n\phi \int_{-\pi}^{\pi} H(\xi) \sin n\xi \, d\xi \right) \right],$$

which, by the uniformity of the convergence, can be written

$$H^{(r)}(\phi) = \frac{1}{2\pi} \int_{-\pi}^{\pi} H(\xi) \, d\xi + \frac{1}{\pi} \int_{-\pi}^{\pi} \left\{ \sum_{n=1}^{\infty} [r^n(\cos n\phi \cos n\xi \right.$$
$$\left. + \sin n\phi \sin n\xi)H(\xi)] \right\} d\xi. \tag{5.14}$$

Simple manipulation next reduces (5.14) to

$$H^{(r)}(\phi) = \frac{1}{\pi} \int_{-\pi}^{\pi} H(\xi) \left\{ \frac{1}{2} + \sum_{n=1}^{\infty} [r^n \cos n(\phi - \xi)] \right\} d\xi. \tag{5.15}$$

However, by Euler's identity and by means of (1.30), it follows that for $0 \le r < 1$,

$$\frac{1}{2} + \sum_{n=1}^{\infty} [r^n \cos n(\phi - \xi)] = \frac{1}{2} + \frac{1}{2} \sum_{1}^{\infty} [r^n e^{in(\phi-\xi)} + r^n e^{-in(\phi-\xi)}]$$

$$= \frac{1}{2} \left[1 + \frac{re^{i(\phi-\xi)}}{1 - re^{i(\phi-\xi)}} + \frac{re^{-i(\phi-\xi)}}{1 - re^{-i(\phi-\xi)}} \right]$$

$$= \frac{1}{2} \left[\frac{1 - r^2}{1 - re^{-i(\phi-\xi)} - re^{i(\phi-\xi)} + r^2} \right]$$

$$= \frac{1}{2} \left[\frac{1 - r^2}{1 - 2r \cos(\phi - \xi) + r^2} \right]$$

so that

$$H^{(r)}(\phi) = \frac{1}{2\pi} \int_{-\pi}^{\pi} \frac{H(\xi)(1 - r^2)}{1 - 2r \cos(\phi - \xi) + r^2} \, d\xi. \tag{5.16}$$

Thus, (5.13) and (5.16) imply

$$\frac{a_0}{2} + \sum_{n=1}^{\infty} \left[r^n (a_n \cos n\phi + b_n \sin n\phi) \right] \equiv \frac{1}{2\pi} \int_{-\pi}^{\pi} \frac{H(\xi)(1 - r^2)}{1 - 2r \cos(\phi - \xi) + r^2} \, d\xi,$$

(5.17)

where a_n, b_n are given by (5.11) and (5.12), $0 \leq r < 1$. For $a_0 = 2$, $a_n = b_n = 0; n = 1, 2, 3, \ldots$, (5.10) implies $H(\phi) \equiv 1$, so that, from (5.17), one has that, independently of ϕ and r,

$$1 \equiv \frac{1}{2\pi} \int_{-\pi}^{\pi} \frac{1 - r^2}{1 - 2r \cos(\phi - \xi) + r^2} \, d\xi.$$

(5.18)

From (5.18), then,

$$H(\phi) \equiv \frac{1}{2\pi} \int_{-\pi}^{\pi} \frac{H(\phi)(1 - r^2)}{1 - 2r \cos(\phi - \xi) + r^2} \, d\xi.$$

(5.19)

Thus, for $0 \leq r < 1$,

$$f(x,y) - H(\phi) = \frac{1}{2\pi} \int_{-\pi}^{\pi} \frac{(1 - r^2)[H(\xi) - H(\phi)]}{1 - 2r \cos(\phi - \xi) + r^2} \, d\xi.$$

(5.20)

Now, since $H(\phi)$ is continuous on $-\pi \leq \phi \leq \pi$, it is uniformly continuous. Thus, given $\epsilon > 0$, there exists a positive number $\delta(\epsilon)$ such that $|\phi - \xi| < \delta$ implies $|H(\phi) - H(\xi)| < \epsilon$. In addition, if $|\phi - \xi| \geq \delta$, then $\phi - \xi \neq 0$, so that

$$\lim_{r \to 1^-} \frac{1 - r^2}{1 - 2r \cos(\phi - \xi) + r^2} = 0,$$

which implies that there exists r^* such that if $0 < r^* \leq r < 1$ and $|\phi - \xi| \geq \delta$, then

$$0 < \frac{1 - r^2}{1 - 2r \cos(\phi - \xi) + r^2} < \epsilon.$$

(5.21)

Hence, for $0 < r^* \leq r < 1$,

$$|f(x,y) - H(\phi)| \leq \frac{1}{2\pi} \int_{\substack{-\pi \\ |\phi - \xi| \geq \delta}}^{\pi} \frac{(1 - r^2)|H(\xi) - H(\phi)|}{1 - 2r \cos(\phi - \xi) + r^2} \, d\xi$$

$$+ \frac{1}{2\pi} \int_{\substack{-\pi \\ |\phi - \xi| < \delta}}^{\pi} \frac{(1 - r^2)|H(\xi) - H(\phi)|}{1 - 2r \cos(\phi - \xi) + r^2} \, d\xi$$

$$\leq \frac{1}{2\pi} [2\pi\epsilon] \left[\max_{0 \leq \phi \leq 2\pi} 2|H(\phi)| \right] + \frac{\epsilon}{2\pi} \int_{\substack{-\pi \\ |\phi - \xi| < \delta}}^{\pi} \frac{1 - r^2}{1 - 2r \cos(\phi - \xi) + r^2} \, d\xi.$$

From (5.18) then

$$|f(x,y) - H(\phi)| \leq \epsilon[1 + 2(\max_{0 \leq \phi \leq 2\pi} |H(\phi)|)], \tag{5.22}$$

so that (5.21) and (5.22) imply that, uniformly in ϕ,

$$\lim_{r \to 1^-} f(x,y) = H(\phi), \qquad x^2 + y^2 < 1.$$

The theorem is thus valid for the case $H(\phi) \in C^2$ on $-\pi \leq \phi \leq \pi$.

Suppose next that $H(\phi)$ is continuous on $-\pi \leq x \leq \pi$ and satisfies $H(-\pi) = H(\pi)$. Then, by Theorem 2.7, $H(\phi)$ can be approximated uniformly by a finite trigonometric sum of the form

$$H_{(n)}(\phi) = \frac{a_0}{2} + \sum_{k=1}^{n} (a_k \cos k\phi + b_k \sin k\phi). \tag{5.23}$$

Note that $H_{(n)}(\phi) \in C^2$ for $-\pi \leq \phi \leq \pi$ and that $H_{(n)}(-\pi) = H_{(n)}(\pi)$. One can then construct a sequence of functions $f_n(x,y)$, $n = 1, 2, \ldots$, defined by

$$f_n(x,y) = H_{(n)}(\phi), \qquad x^2 + y^2 = 1, \tag{5.24}$$

$$f_n(x,y) = \frac{1}{2\pi} \int_{-\pi}^{\pi} \frac{H_{(n)}(\xi)(1 - x^2 - y^2)}{1 - 2x \cos \xi - 2y \sin \xi + x^2 + y^2} \, d\xi, \tag{5.25}$$

such that for each n, $f_n(x,y)$ is the solution of the Dirichlet problem with $g(x,y) = H_{(n)}(\phi)$. Since the sequence $\{H_{(n)}(\phi)\}$ converges uniformly on B to $H(\phi)$, it follows from Theorem 5.7 that the sequence $\{f_n(x,y)\}$ converges uniformly on $\bar{R}$. Hence, taking limits in (5.24) and (5.25), one has $\lim_{n \to \infty} f_n(x,y) = f(x,y)$, where

$$f(x,y) = g(x,y) \equiv H(\phi); \qquad x^2 + y^2 = 1; \tag{5.26}$$

$$f(x,y) = \frac{1}{2\pi} \int_{-\pi}^{\pi} \frac{H(\xi)(1 - x^2 - y^2)}{1 - 2x \cos \xi - 2y \sin \xi + x^2 + y^2} \, d\xi; \qquad x^2 + y^2 < 1. \tag{5.27}$$

Direct calculation shows (5.27) to be harmonic on R while $f(x,y)$ is continuous on $\bar{R}$ since it is the limit of a uniformly convergent sequence of continuous functions on a closed and bounded set. Thus $f(x,y)$, defined by (5.26) and (5.27), is the desired solution and the theorem is proved.

Example. If R is the interior of the unit circle, B is the unit circle, and $g(x,y) = 1 + x^2 - y$ on B, find the solution of the Dirichlet problem on $\bar{R} = R \cup B$.

Solution. Let B be parametrized by $x = \cos\phi, y = \sin\phi, -\pi \leq \phi \leq \pi$. Then let $g(x,y) = 1 + x^2 - y = 1 + \cos^2\phi - \sin\phi = H(\phi)$. Then by Theorem 5.8, the unique solution is

$$f(x,y) = \begin{cases} 1 + x^2 - y, & x^2 + y^2 = 1, \\ \dfrac{1}{2\pi}\displaystyle\int_{-\pi}^{\pi} \dfrac{(1 + \cos^2\xi - \sin\xi)(1 - x^2 - y^2)}{1 - 2x\cos\xi - 2y\sin\xi + x^2 + y^2}\, d\xi, & x^2 + y^2 < 1. \end{cases}$$

Theorem 5.9. For the Dirichlet problem, let R be the interior of the unit circle, and let its boundary B be given parametrically by $x = \cos\phi, y = \sin\phi, -\pi \leq \phi \leq \pi$. Let $g(x,y) = g(\cos\phi, \sin\phi) = H(\phi)$. If $u = f(x,y)$ is defined on $\bar{R} = R \cup B$ by

$$f(x,y) = g(x,y) \equiv H(\phi), \qquad x^2 + y^2 = 1, \qquad (5.28)$$

$$f(x,y) = \frac{a_0}{2} + \sum_{n=1}^{\infty} [r^n(a_n \cos n\phi + b_n \sin n\phi)], \qquad x^2 + y^2 = r^2 < 1,$$
$$(5.29)$$

where (r,ϕ) are polar coordinates of (x,y) and

$$a_n = \frac{1}{\pi}\int_{-\pi}^{\pi} H(\xi)\cos n\xi\, d\xi, \qquad b_n = \frac{1}{\pi}\int_{-\pi}^{\pi} H(\xi)\sin n\xi\, d\xi.$$

Then the function so defined is the solution of the given Dirichlet problem.

Proof. The proof follows readily from Theorem 5.8 and identity (5.17).

Theorem 5.10. For the Dirichlet problem, let B be the circle whose equation is $x^2 + y^2 = \rho^2, \rho > 0$, and let R be the interior of B. Let B be given parametrically by $x = \rho\cos\phi, y = \rho\sin\phi, -\pi \leq \phi \leq \pi$, so that $g(x,y) = g(\rho\cos\phi, \rho\sin\phi) = H(\phi)$. If (r,ϕ) are polar coordinates of (x,y), then $u = f(x,y)$, defined by

$$f(x,y) = g(x,y) \equiv H(\phi), \qquad x^2 + y^2 = \rho^2, \qquad (5.30)$$

$$f(x,y) = \frac{1}{2\pi}\int_{-\pi}^{\pi} \frac{(\rho^2 - x^2 - y^2)H(\xi)}{\rho^2 - 2\rho x\cos\xi - 2\rho y\sin\xi + x^2 + y^2}\, d\xi,$$
$$x^2 + y^2 < \rho^2 \quad (5.31)$$

is the unique solution of the Dirichlet problem under consideration.

Proof. The proof follows essentially by replacing r in Poisson integral (5.8′) by r/ρ, simplifying, and then expressing the resulting integral in rectangular cartesian coordinates.

Of course, by means of polar coordinates, (5.31), like (5.8), can be expressed in an alternate form, namely,

$$f(x,y) = \frac{1}{2\pi} \int_{-\pi}^{\pi} \frac{(\rho^2 - r^2)H(\xi)}{\rho^2 - 2\rho r \cos(\phi - \xi) + r^2} \, d\xi. \qquad (5.31')$$

Example 1.   Let B be the circle whose equation is $x^2 + y^2 = 4$ and let R be the interior of B.   Let $g(x,y) = 1 + x^2 - y$ on B.   Then the solution of the corresponding Dirichlet problem is

$$f(x,y) = \begin{cases} 1 + x^2 - y, & x^2 + y^2 = 4, \\ \dfrac{1}{2\pi} \displaystyle\int_{-\pi}^{\pi} \dfrac{(4 - x^2 - y^2)(1 + 4\cos^2 \xi - 2\sin \xi)}{4 - 4x\cos \xi - 4y\sin \xi + x^2 + y^2} d\xi, & x^2 + y^2 < 4, \end{cases}$$

or, equivalently, in terms of polar coordinates,

$$f(x,y) = \begin{cases} 1 + 4\cos^2 \phi - 2\sin \phi, & x^2 + y^2 = r^2 = 4, \\ \dfrac{1}{2\pi} \displaystyle\int_{-\pi}^{\pi} \dfrac{(4 - r^2)(1 + 4\cos^2 \xi - 2\sin \xi)}{4 - 4r\cos(\phi - \xi) + r^2} d\xi, & x^2 + y^2 = r^2 < 4. \end{cases}$$

Example 2.   Let B be the circle whose equation is $(x - 1)^2 + (y + 2)^2 = 4$ and let R be the interior of B.   Let $g(x,y) = 1 + (x - 1)^2 - (y + 2)$ on B.   Then, setting $\bar{x} = x - 1$, $\bar{y} = y + 2$, the solution of the corresponding Dirichlet problem is

$$f(\bar{x},\bar{y}) = \begin{cases} 1 + \bar{x}^2 - \bar{y}, & \bar{x}^2 + \bar{y}^2 = 4, \\ \dfrac{1}{2\pi} \displaystyle\int_{-\pi}^{\pi} \dfrac{(4 - \bar{x}^2 - \bar{y}^2)(1 + 4\cos^2 \xi - 2\sin \xi)}{4 - 4\bar{x}\cos \xi - 4\bar{y}\sin \xi + \bar{x}^2 + \bar{y}^2} d\xi, & \bar{x}^2 + \bar{y}^2 < 4. \end{cases}$$

The continuity of $f(\bar{x},\bar{y})$ follows from the fact that a continuous function of a continuous function is continuous while the harmonic property is a direct consequence of Theorem 5.1.

Theorem 5.11.   For the Dirichlet problem, let B be the circle whose equation is $x^2 + y^2 = \rho^2$, $\rho > 0$, and let R be the interior of B.   Let B be given parametrically by $x = \rho\cos \phi$, $y = \rho\sin \phi$, $-\pi \le \phi \le \pi$, so that $g(x,y) = g(\rho\cos \phi, \rho\sin \phi) = H(\phi)$.   If (r,ϕ) are polar coordinates of (x,y), then $u = f(x,y)$, defined by

$$f(x,y) = g(x,y) = H(\phi), \qquad x^2 + y^2 = \rho^2, \qquad (5.32)$$

$$f(x,y) = \frac{a_0}{2} + \sum_{n=1}^{\infty} \left[\left(\frac{r}{\rho}\right)^n (a_n \cos n\phi + b_n \sin n\phi) \right], \qquad x^2 + y^2 = r^2 < \rho^2,$$

$$(5.33)$$

where

$$a_n = \frac{1}{\pi} \int_{-\pi}^{\pi} H(\xi) \cos n\xi \, d\xi, \qquad b_n = \frac{1}{\pi} \int_{-\pi}^{\pi} H(\xi) \sin n\xi \, d\xi,$$

is the solution of the Dirichlet problem under consideration.

Proof. The proof follows in a fashion analogous to that of Theorem 5.9.

Theorem 5.12. For the Dirichlet problem, let B be the circle whose equation is $x^2 + y^2 = \rho^2$, $\rho > 0$, and let R be the interior of B. Let B be given parametrically by $x = \rho \cos \phi$, $y = \rho \sin \phi$, $-\pi \leq \phi \leq \pi$. If $f_1(x,y)$ is the solution which corresponds to boundary function $g(x,y) = g_1(x,y)$ and $f_2(x,y)$ is the solution which corresponds to boundary function $g(x,y) = g_2(x,y)$, and if on B $g_1(x,y) \leq g_2(x,y)$, then $f_1(x,y) \leq f_2(x,y)$ on $R \cup B$.

Proof. Let $g_1(x,y) = g_1(\rho \cos \phi, \rho \sin \phi) = H_1(\phi)$; $g_2(x,y) = g_2(\rho \cos \phi, \rho \sin \phi) = H_2(\phi)$. Then, on R,

$$f_1(x,y) = \frac{1}{2\pi} \int_{-\pi}^{\pi} \frac{(\rho^2 - x^2 - y^2)H_1(\xi)}{\rho^2 - 2\rho x \cos \xi - 2\rho y \sin \xi + x^2 + y^2} d\xi, \qquad x^2 + y^2 < \rho^2,$$

$$f_2(x,y) = \frac{1}{2\pi} \int_{-\pi}^{\pi} \frac{(\rho^2 - x^2 - y^2)H_2(\xi)}{\rho^2 - 2\rho x \cos \xi - 2\rho y \sin \xi + x^2 + y^2} d\xi, \qquad x^2 + y^2 < \rho^2.$$

However, as shown in Theorem 5.8, the denominator of the Poisson integral is positive. Since $[\rho^2 - x^2 - y^2]$ is positive on R and since $H_1 \leq H_2$ on B, it follows that on R

$$f_1(x,y) - f_2(x,y) = \frac{1}{2\pi} \int_{-\pi}^{\pi} \frac{(\rho^2 - x^2 - y^2)[H_1(\xi) - H_2(\xi)]}{\rho^2 - 2\rho x \cos \xi - 2\rho y \sin \xi + x^2 + y^2} d\xi \leq 0,$$

or, equivalently, that $f_1(x,y) \leq f_2(x,y)$. Since $f_1(x,y) \equiv g_1(x,y)$, $f_2(x,y) \equiv g_2(x,y)$ on B, and $g_1(x,y) \leq g_2(x,y)$ on B, it follows that $f_1(x,y) \leq f_2(x,y)$ on $R \cup B$, and the theorem is proved.

5.3   Further Properties of Harmonic Functions

With the aid of the Poisson integral and its immediate consequences, one can develop further properties of harmonic functions which will be of great value in establishing the existence of a unique solution of the general Dirichlet problem.

Theorem 5.13. If $u = f(x,y)$ is continuous and harmonic on region R, then $f(x,y)$ is analytic at each point of R.

Proof. Since the harmonic property of a given function is invariant under translation of axes, assume without loss of generality that the point $(0,0) \in R$. It is only necessary then to show that $u = f(x,y)$ is analytic at $(0,0)$.

Since every point of a region is an interior point, there exists circular region R_1 with boundary B_1, radius r_1, and center $(0,0)$ such that $(R_1 \cup B_1) \subset R$. Then, if B_1 is parametrized by $x = r_1 \cos \phi$, $y = r_1 \sin \phi$, $-\pi \le \phi \le \pi$, Theorem 5.10 implies

$$f(x,y) = \frac{a_0}{2} + \sum_{n=1}^{\infty} \left[\left(\frac{r}{r_1} \right)^n (a_n \cos n\phi + b_n \sin n\phi) \right], \qquad 0 \le r < r_1.$$

(5.34)

Rearrangement of (5.34) yields

$$f(x,y) = \frac{a_0}{2} + \sum_{n=1}^{\infty} \left[\frac{a_n}{(r_1)^n} r^n \cos n\phi + \frac{b_n}{(r_1)^n} r^n \sin n\phi \right]. \quad (5.35)$$

However, by DeMoivre's theorem,

$$(\cos \phi + i \sin \phi)^n \equiv \cos n\phi + i \sin n\phi. \qquad (5.36)$$

Expanding the left side of (5.36) by the binomial theorem and then setting $\cos n\phi$ equal to the real part of that expansion and $\sin n\phi$ equal to the imaginary part yields that

$$\cos n\phi = \sum_{k=0}^{n} A_k'(\cos \phi)^k(\sin \phi)^{n-k}, \qquad \sin n\phi = \sum_{k=0}^{n} A_k''(\cos \phi)^k(\sin \phi)^{n-k},$$

where A_k', A_k'', except for sign, are binomial coefficients. Hence

$$r^n \cos n\phi = \sum_{k=0}^{n} A_k'(r \cos \phi)^k(r \sin \phi)^{n-k} = \sum_{k=0}^{n} A_k' x^k y^{n-k}, \qquad (5.37)$$

$$r^n \sin n\phi = \sum_{k=0}^{n} A_k''(r \cos \phi)^k(r \sin \phi)^{n-k} = \sum_{k=0}^{n} A_k'' x^k y^{n-k}. \qquad (5.38)$$

Substitution of (5.37) and (5.38) into (5.35) then readily implies the validity of the theorem.

The introduction now of the concept of *mean value* will lead to a most remarkable theorem which describes an alternate way of viewing harmonic functions.

Definition 5.4. Let $f(x,y)$ be defined on region R. Let R_1 be a circular region with boundary B_1, radius r_1, and center (x_1,y_1) such that $(R_1 \cup B_1) \subset R$. If, for each possible such R_1,

$$f(x_1,y_1) = \frac{1}{2\pi r_1} \oint_{B_1} f(x,y) \, ds,$$

then $f(x,y)$ is said to possess the *mean value property* on R.

Theorem 5.14. Let R be a simply connected, bounded region whose boundary B is a contour. If $u = f(x,y)$ is continuous on $\bar{R} = R \cup B$ and possesses the mean value property on R, then $u = f(x,y)$ assumes its maximum value on B and its minimum value on B.

Proof. If $R^* \subset \bar{R}$ is the set of points on which $u = f(x,y)$ assumes its maximum value, then the continuity of $f(x,y)$ implies that R^* is closed. Since $R^* \subset \bar{R}$, R^* is bounded. Let d be the distance between the two closed, bounded sets R^* and B. If $d = 0$, the theorem is valid. Suppose then $d \neq 0$. Then $f(x,y)$ assumes its maximum at some point $(x_1,y_1) \in R^*$ such that the distance from (x_1,y_1) to B is equal to d. Let R_1 be the circular region with boundary B_1, radius $d_1 = d/2$, and center (x_1,y_1). Then, by the mean value property,

$$f(x_1,y_1) = \frac{1}{2\pi d_1} \oint_{B_1} f(x,y)\, ds. \tag{5.39}$$

But since B_1 contains points which are not in R^* and $f(x_1,y_1)$ is the maximum value $f(x,y)$ can assume, (5.39) cannot be valid, and the maximum property follows by contradiction. The minimum property then follows by applying the maximum property to $-f(x,y)$.

Theorem 5.15. Let R be a simply connected, bounded region whose boundary B is a contour. If $f_1(x,y)$, $f_2(x,y)$ are continuous on $\bar{R} = R \cup B$, have the mean value property on R, and coincide on B, then $f_1(x,y) \equiv f_2(x,y)$ on $\bar{R}$.

Proof. From Definition 5.4, it follows that the function

$$f(x,y) = f_1(x,y) - f_2(x,y)$$

has the mean value property on R. Moreover, $f(x,y)$ is continuous on $\bar{R}$. Then $f(x,y)$ assumes its maximum and minimum values on B. But $f_1(x,y) - f_2(x,y) \equiv 0$ on B. Thus, $f_1(x,y) - f_2(x,y) \equiv 0$ on $\bar{R}$ and the theorem follows readily.

Theorem 5.16. Let R be a simply connected, bounded region whose boundary B is a contour. If $u = f(x,y)$ is continuous on $\bar{R} = R \cup B$, then a necessary and sufficient condition that $u = f(x,y)$ be harmonic on R is that it possess the mean value property on R.

Proof. Suppose first that $f(x,y)$ is harmonic on R. Since the harmonic property is invariant under translation, assume without loss of generality that $(0,0) \in R$. It is only necessary to consider circular region R_1, with boundary B_1, radius r_1, and center $(0,0)$ such that $(R_1 \cup B_1) \subset R$. Let B_1 have parametric representation $x = r_1 \cos \phi$, $y = r_1 \sin \phi$, $-\pi \leq \phi \leq \pi$, so that for $(x,y) \in B_1$, $f(x,y) = f(r_1 \cos \phi, r_1 \sin \phi) = H(\phi)$. Thus, for all $0 \leq r_2 < r_1$, (5.31') implies

$$f(r_2 \cos \phi, r_2 \sin \phi) = \frac{1}{2\pi} \int_{-\pi}^{\pi} \frac{(r_1^2 - r_2^2)H(\xi)}{r_1^2 - 2r_1 r_2 \cos(\phi - \xi) + r_2^2}\, d\xi. \tag{5.40}$$

Hence, for $r_2 = 0$, (5.40) implies

$$f(0,0) = \frac{1}{2\pi} \int_{-\pi}^{\pi} H(\xi)\, d\xi = \frac{1}{2\pi r_1} \oint_{B_1} f(x,y)\, ds,$$

and the necessity is proved.

Suppose next that $u = f(x,y)$ has the mean value property on R. Then let $(x_1,y_1) \in R$ and let R_1 be a circular region with boundary B_1, radius r_1, center (x_1,y_1), such that $(R_1 \cup B_1) \subset R$. Then, by Theorem 5.10, there exists a unique function $f^*(x,y)$ defined on $R_1 \cup B_1$ such that $f^*(x,y)$ is harmonic on R_1, continuous on $R_1 \cup B_1$, and identical with $f(x,y)$ on B_1. Moreover, by the proof of the necessity, $f^*(x,y)$ has the mean value property on R_1. Thus, on $R_1 \cup B_1$, $f^*(x,y) \equiv f(x,y)$, so that $f(x,y)$ is harmonic on R_1. Since (x_1,y_1) was an arbitrary point of R, it follows that $f(x,y)$ is harmonic on R. Thus the sufficiency is established and the theorem is proved.

Another theorem on harmonic functions with extensive application is Harnack's theorem.

Theorem 5.17.   Harnack's Theorem.   Let R be a simply connected, bounded region whose boundary B is a contour. Set $\bar{R} = R \cup B$. Let $\{f_n(x,y)\}$ be a sequence of functions each of which is continuous on $\bar{R}$ and harmonic on R. If $\{f_n(x,y)\}$ converges uniformly on B, then $\{f_n(x,y)\}$ converges on $\bar{R}$ to a limit function which is continuous on $\bar{R}$ and harmonic on R.

Proof.   By Theorem 5.7, $\{f_n(x,y)\}$ converges uniformly on $\bar{R}$. Since on a closed, bounded set a uniformly convergent sequence of continuous functions converges to a function which is continuous on that set, then $\{f_n(x,y)\}$ converges to a function $f(x,y)$ which is continuous on $\bar{R}$. In order to show that $f(x,y)$ is harmonic on R, let $(x,y) \in R$. Since all points of R are interior points, one can represent each $f_n(x,y)$ by a Poisson integral of the form

$$f_n(x,y) = \frac{1}{2\pi} \int_{-\pi}^{\pi} \frac{(\rho^2 - x^2 - y^2)H_n(\xi)}{\rho^2 - 2\rho x \cos \xi - 2\rho y \sin \xi + x^2 + y^2}\, d\xi, \qquad x^2 + y^2 < \rho^2,$$

where the circle with center (x,y) and radius ρ is, with its interior, contained in R. By the uniformity of the convergence,

$$f(x,y) = \lim_{n \to \infty} f_n(x,y)$$

$$= \frac{1}{2\pi} \int_{-\pi}^{\pi} \left[\lim_{n \to \infty} \frac{(\rho^2 - x^2 - y^2)H_n(\xi)}{\rho^2 - 2\rho x \cos \xi - 2\rho y \sin \xi + x^2 + y^2} \right] d\xi$$

$$= \frac{1}{2\pi} \int_{-\pi}^{\pi} \frac{(\rho^2 - x^2 - y^2)H(\xi)}{\rho^2 - 2\rho x \cos \xi - 2\rho y \sin \xi + x^2 + y^2}\, d\xi,$$

from which the theorem readily follows.

Also of value will be a result about derivatives of harmonic functions.

Theorem 5.18. Let R be a simply connected, bounded region and let its boundary B be a contour. Let $u = f(x,y)$ be harmonic and uniformly bounded on R, so that there exists a constant c^* such that if $(x,y) \in R$,

$$|f(x,y)| \le c^*.$$

Then on any closed subset $R^* \subset R$, there exists a uniform bound for $\dfrac{\partial f}{\partial x}, \dfrac{\partial f}{\partial y}$ of the form

$$\left| \frac{\partial f}{\partial x} \right| \le \frac{4c^*}{\pi\rho}, \qquad \left| \frac{\partial f}{\partial y} \right| \le \frac{4c^*}{\pi\rho}, \tag{5.41}$$

where 2ρ is the distance between the two closed, bounded sets R^* and B.

Proof. Let $(x,y) \in R^*$. Since $(x,y) \in R$, there exists a circular region R_1 with boundary B_1, radius r_1, center (x,y) such that $(R_1 \cup B_1) \subset R$. By the mean value property, if B_2 is any circular region concentric with B_1 and with radius r such that $0 < r \le r_1$, then

$$f(x,y) = \frac{1}{2\pi r} \oint_{B_2} f(x,y)\, ds = \frac{1}{2\pi} \int_{-\pi}^{\pi} f(x + r\cos\phi, y + r\sin\phi)\, d\phi.$$

Hence,

$$\int_0^{r_1} r f(x,y)\, dr = \frac{1}{2\pi} \int_0^{r_1} \int_{-\pi}^{\pi} f(x + r\cos\phi, y + r\sin\phi) r\, d\phi\, dr.$$

Setting $\xi = x + r\cos\phi, \eta = y + r\sin\phi$ implies

$$\int_0^{r_1} r f(x,y)\, dr = \frac{1}{2\pi} \iint_{(\xi-x)^2+(\eta-y)^2 \le r_1^2} f(\xi,\eta)\, d\xi\, d\eta,$$

so that

$$\frac{r_1^2}{2} f(x,y) = \frac{1}{2\pi} \iint_{(\xi-x)^2+(\eta-y)^2 \le r_1^2} f(\xi,\eta)\, d\xi\, d\eta.$$

Hence

$$f(x,y) = \frac{1}{\pi r_1^2} \iint_{(\xi-x)^2+(\eta-y)^2 \le r_1^2} f(\xi,\eta)\, d\xi\, d\eta$$

$$= \frac{1}{\pi r_1^2} \int_{y-r_1}^{y+r_1} \int_{x-\sqrt{r_1^2-(\eta-y)^2}}^{x+\sqrt{r_1^2-(\eta-y)^2}} f(\xi,\eta)\, d\xi\, d\eta.$$

By the fundamental theorem of calculus,

$$\frac{\partial f(x,y)}{\partial x} = \frac{1}{\pi r_1^2} \int_{y-r_1}^{y+r_1} [f(x + \sqrt{r_1^2 - (\eta - y)^2}, \eta)$$

$$- f(x - \sqrt{r_1^2 - (\eta - y)^2}, \eta)]\, d\eta,$$

so that

$$\left| \frac{\partial f(x,y)}{\partial x} \right| \leq \frac{1}{\pi r_1^2} 2r_1 \cdot 2c^* = \frac{4c^*}{\pi r_1}.$$

Since one can always select r_1 to satisfy $0 < \rho \leq r_1$, and since (x,y) is arbitrary, the above discussion implies that on R^*

$$\left| \frac{\partial f(x,y)}{\partial x} \right| \leq \frac{4c^*}{\pi \rho}. \qquad (5.42)$$

A similar proof holds for $\left| \dfrac{\partial f(x,y)}{\partial y} \right|$ and the theorem is valid.

5.4   On the Dirichlet Problem

The Dirichlet problem has been resolved thus far only for circular regions. However, with the aid of these results it will now be shown that *the unique solution of the Dirichlet problem always exists.* The ingenious method to be used was devised by O. Perron.

Let R be a simply connected, bounded region whose boundary B is a contour. Let R' be a circular region with circular boundary B' such that $R' \subset R$. Suppose also that $w(x,y)$ is continuous on $R \cup B$ and that $u = f(x,y)$ is the unique function which is harmonic on R', continuous on $R' \cup B'$, and identical with $w(x,y)$ on B'. Then the symbol $M_{R'}(w)$ will represent the function defined on $R \cup B$ by

$$M_{R'}[w(x,y)] = \begin{cases} f(x,y), & (x,y) \in R' \\ w(x,y), & (x,y) \in [(R \cup B) \cap (E^2 - R')]. \end{cases} \qquad (5.43)$$

Theorem 5.19. Let R be a simply connected, bounded region whose boundary B is a contour. Let R' be a circular region with boundary B' such that $R' \subset R$. Suppose that $w_1(x,y)$, $w_2(x,y)$ are continuous on $R \cup B$ and that $f_1(x,y)$, $f_2(x,y)$ are harmonic on R', continuous on $R' \cup B'$, and identical, respectively, with $w_1(x,y)$, $w_2(x,y)$ on B'. Then

$$M_{R'}[w_1(x,y) + w_2(x,y)] = M_{R'}[w_1(x,y)] + M_{R'}[w_2(x,y)]. \qquad (5.44)$$

Proof. From (5.43)

$$M_{R'}[w_1] + M_{R'}[w_2] = \begin{cases} f_1(x,y) + f_2(x,y), & (x,y) \in R', \\ w_1(x,y) + w_2(x,y), \\ & (x,y) \in [(R \cup B) \cap (E^2 - R')]. \end{cases}$$
(5.45)

Moreover, $[f_1(x,y) + f_2(x,y)]$ is harmonic on R', continuous on $R' \cup B'$ and identical with $[w_1(x,y) + w_2(x,y)]$ on B' so that

$$M_{R'}[w_1 + w_2] = \begin{cases} f_1(x,y) + f_2(x,y), & (x,y) \in R', \\ w_1(x,y) + w_2(x,y), & (x,y) \in [(R \cup B) \cap (E^2 - R')]. \end{cases}$$
(5.46)

The theorem then follows immediately from (5.45) and (5.46).

Theorem 5.20. Let R be a simply connected, bounded region whose boundary B is a contour. Let R' be a circular region with boundary B', such that $R' \subset R$. Suppose that $w_1(x,y)$, $w_2(x,y)$ are continuous on $R \cup B$ and that $u_1 = f_1(x,y)$, $u_2 = f_2(x,y)$ are harmonic on R', continuous on $R' \cup B'$, and identical, respectively, with $w_1(x,y)$, $w_2(x,y)$ on B'. Then if $w_1(x,y) \leq w_2(x,y)$ on $R \cup B$, $M_{R'}[w_1(x,y)] \leq M_{R'}[w_2(x,y)]$.

Proof. Since

$$M_{R'}[w_1(x,y)] = \begin{cases} f_1(x,y), & (x,y) \in R', \\ w_1(x,y), & (x,y) \in [(R \cup B) \cap (E^2 - R')], \end{cases}$$
(5.47)

$$M_{R'}[w_2(x,y)] = \begin{cases} f_2(x,y), & (x,y) \in R' \\ w_2(x,y), & (x,y) \in [(R \cup B) \cap (E^2 - R')], \end{cases}$$
(5.48)

the theorem then will be valid if $f_1(x,y) \leq f_2(x,y)$ on R'. This is, however, a direct consequence of Theorem 5.12 and the theorem is established.

Definition 5.5. Let R be a simply connected, bounded region whose boundary B is a contour. If, for every possible choice of circular region R' with boundary B' such that $R' \subset R$, a given function $w(x,y)$ which is continuous on $R \cup B$ satisfies the relationship

$$w(x,y) \leq M_{R'}[w(x,y)], \qquad (x,y) \in (R \cup B), \tag{5.49}$$

then $w(x,y)$ is said to be subharmonic on $R \cup B$.

Definition 5.6. Let R be a simply connected, bounded region whose boundary B is a contour. If, for every possible choice of circular region R' with boundary B' such that $R' \subset R$, a given function $w(x,y)$ which is continuous on $R \cup B$ satisfies the relationship

$$w(x,y) \geq M_{R'}[w(x,y)], \qquad (x,y) \in (R \cup B), \tag{5.50}$$

then $w(x,y)$ is said to be superharmonic on $R \cup B$.

Theorem 5.21. Every harmonic function is both subharmonic and superharmonic.

Proof. The proof is a direct consequence of Definitions 5.5 and 5.6.

Theorem 5.22. Let R be a simply connected, bounded region whose boundary B is a contour. If $w_1(x,y)$, $w_2(x,y)$ are subharmonic on $R \cup B$, then $w(x,y) = w_1(x,y) + w_2(x,y)$ is subharmonic on $R \cup B$.

Proof. If R' is any circular region with boundary B' such that $R' \subset R$, then

$$\left. \begin{aligned} w_1(x,y) &\leq M_{R'}[w_1(x,y)] \\ w_2(x,y) &\leq M_{R'}[w_2(x,y)]. \end{aligned} \right\} \tag{5.51}$$

By Definition 5.5 and (5.44) then

$$\begin{aligned} w(x,y) &= w_1(x,y) + w_2(x,y) \\ &\leq M_{R'}[w_1] + M_{R'}[w_2] \\ &= M_{R'}[w_1 + w_2] \\ &= M_{R'}[w(x,y)], \end{aligned}$$

and the theorem follows readily.

Theorem 5.23. Let R be a simply connected, bounded region whose boundary B is a contour. If $w_1(x,y)$, $w_2(x,y)$, $w_3(x,y)$, $\ldots$, $w_n(x,y)$ are subharmonic on $R \cup B$, then $w(x,y) = \sum_{i=1}^{n} w_i(x,y)$ is subharmonic on $R \cup B$.

Proof. The proof follows directly from Theorem 5.22.

Theorem 5.24. Let R be a simply connected, bounded region whose boundary B is a contour. If $w_1(x,y)$, $w_2(x,y)$, $\ldots$, $w_n(x,y)$ are subharmonic on $R \cup B$, then

$$w(x,y) = \max\left[w_1(x,y), w_2(x,y), \ldots, w_n(x,y)\right], \qquad (x,y) \in (R \cup B)$$

is subharmonic on $R \cup B$.

Proof. Let R' be any circular region with boundary B' such that $R' \subset R$. Then by Definition 5.5 and Theorem 5.20,

$$w_j(x,y) \leq M_{R'}[w_j(x,y)] \leq M_{R'}[w(x,y)]; \qquad j = 1, 2, \ldots, n.$$

Since for any (x,y), $w(x,y)$ is equal to at least one of the w_j's, it follows immediately that

$$w(x,y) \leq M_{R'}[w(x,y)],$$

which implies the validity of the theorem.

Theorem 5.25. Let R be a simply connected, bounded region whose boundary B is a contour. Let $w(x,y)$ be subharmonic on $R \cup B$. Then $w(x,y)$ assumes its maximum value on B.

Proof. Let R^* be the set of points of $R \cup B$ at which $w(x,y)$ assumes its maximum value W. The continuity of $w(x,y)$ implies that R^* is closed. Since $R^* \subset (R \cup B)$, R^* is bounded. Let d be the distance between the two closed, bounded sets R^* and B. If $d = 0$, the theorem is valid. Suppose $d \neq 0$. Then $w(x,y)$ assumes its maximum value W at some point $(x',y') \in R^*$ such that the distance from (x',y') to B is equal to d. Let R' be the circular region with boundary B', radius $r' = \frac{1}{2}d$, and center (x',y'). Then, by (5.43) and Theorem 5.16,

$$w(x',y') \leq M_{R'}[w(x',y')] = f(x',y')$$

$$= \frac{1}{2\pi r'} \oint_{B'} f(x,y)\, ds = \frac{1}{2\pi r'} \oint_{B'} w(x,y)\, ds,$$

so that

$$w(x',y') \leq \frac{1}{2\pi r'} \oint_{B'} w(x,y)\, ds. \tag{5.52}$$

However, (5.52) is not possible since $w(x',y') = W$ and B' contains points which are not elements of R^*. Thus d must equal zero and the theorem is proved.

Theorem 5.26. Let R be a simply connected, bounded region whose boundary B is a contour. Then $w(x,y)$ is subharmonic on $R \cup B$ if and only if $w^*(x,y) = -w(x,y)$ is superharmonic on $R \cup B$.

Proof. Let R' be a circular region with boundary B' such that $R' \subset R$. Then if $w(x,y)$ is subharmonic on $R \cup B$,

$$w^*(x,y) = -w(x,y) \geq -M_{R'}[w(x,y)] = M_{R'}[-w(x,y)] = M_{R'}[w^*(x,y)],$$

so that $w^*(x,y)$ is superharmonic on $R \cup B$. Conversely, if w^* is superharmonic on $R \cup B$, then

$$w(x,y) = -w^*(x,y) \leq -M_{R'}[w^*(x,y)] = M_{R'}[-w^*(x,y)] = M_{R'}[w(x,y)],$$

so that $w(x,y)$ is subharmonic on $R \cup B$, and the theorem is proved.

Theorem 5.27. Let R be a simply connected, bounded region whose boundary B is a contour. If $w_1(x,y)$, $w_2(x,y)$, ..., $w_n(x,y)$ are superharmonic on $R \cup B$, then $w(x,y) = \sum_{i=1}^{n} w_i(x,y)$ is superharmonic on $R \cup B$.

Proof. The proof follows directly from Theorems 5.23 and 5.26.

Theorem 5.28. Let R be a simply connected, bounded region whose boundary B is a contour. If $w_1(x,y)$, $w_2(x,y)$, . . . , $w_n(x,y)$ are superharmonic on $R \cup B$, then

$$w(x,y) = \min [w_1(x,y), w_2(x,y), \ldots , w_n(x,y)], \qquad (x,y) \in (R \cup B)$$

is superharmonic on $R \cup B$.

Proof. The proof follows directly from Theorems 5.24 and 5.26.

Theorem 5.29. Let R be a simply connected, bounded region whose boundary B is a contour. Let $w(x,y)$ be superharmonic on $R \cup B$. Then $w(x,y)$ assumes its minimum value on B.

Proof. The proof follows directly from Theorems 5.25 and 5.26.

Definition 5.7. Let R be a simply connected, bounded region whose boundary B is a contour. Let $g(x,y)$ be defined and continuous on B. If:

(a) $w(x,y)$ is continuous on $R \cup B$,
(b) $w(x,y) \le g(x,y)$ on B,
(c) $w(x,y)$ is subharmonic on $R \cup B$,

then $w(x,y)$ is said to be a *subfunction* of $g(x,y)$ on $R \cup B$.

Definition 5.8. Let R be a simply connected, bounded region whose boundary B is a contour. Let $g(x,y)$ be defined and continuous on B. If:

(a) $w(x,y)$ is continuous on $R \cup B$,
(b) $w(x,y) \ge g(x,y)$ on B,
(c) $w(x,y)$ is superharmonic on $R \cup B$,

then $w(x,y)$ is said to be a *superfunction* of $g(x,y)$ on $R \cup B$.

The existence of sub- and superfunctions follows from a consideration of $c = \max_B |g(x,y)|$. For $w(x,y) = c$ is harmonic and hence, by Theorem 5.21, both subharmonic and superharmonic on $R \cup B$. Moreover, $c \ge g(x,y)$ on B and $w(x,y) = c$ is continuous on $R \cup B$. Thus, $w(x,y) = c$ is a superfunction of $g(x,y)$ on $R \cup B$. In an analogous fashion, it follows readily that $w(x,y) = -c$ is a subfunction of $g(x,y)$ on $R \cup B$.

Theorem 5.30. Let R be a simply connected, bounded region whose boundary B is a contour. Let $g(x,y)$ be defined and continuous on B and let $c = \max_B |g(x,y)|$. If $w(x,y)$ is any subfunction of $g(x,y)$ on $R \cup B$, then $w(x,y) \le c$ on $R \cup B$.

Proof. The proof is a direct consequence of Theorem 5.25 and the fact that $w(x,y) \le g(x,y)$ on B.

Theorem 5.31. Let R be a simply connected, bounded region whose boundary B is a contour. Let $g(x,y)$ be defined and continuous on B and let $c = \max_{B} |g(x,y)|$. If $w(x,y)$ is any superfunction of $g(x,y)$ on $R \cup B$, then $w(x,y) \geq -c$ on $R \cup B$.

Proof. The proof is a direct consequence of Theorem 5.29 and the fact that $w(x,y) \geq g(x,y)$ on B.

Theorem 5.32. Let R be a simply connected, bounded region whose boundary B is a contour. Let $g(x,y)$ be defined and continuous on B. If $g_1(x,y), g_2(x,y), \ldots, g_n(x,y)$ are superfunctions of $g(x,y)$ on $R \cup B$ and $\sum_{i=1}^{n} g_i(x,y) \geq g(x,y)$ on B, then

$$w(x,y) = \sum_{i=1}^{n} [g_i(x,y)] \tag{5.53}$$

is a superfunction of $g(x,y)$ on $R \cup B$.

Proof. By Theorem 5.27, $w(x,y)$ is superharmonic on $R \cup B$. Since $g_1(x,y), g_2(x,y), \ldots, g_n(x,y)$ are continuous on $R \cup B$, so is $w(x,y)$. Moreover, by assumption, $w(x,y) \geq g(x,y)$ on B. Hence, $w(x,y)$, given by (5.53), is a superfunction of $g(x,y)$ on $R \cup B$.

Theorem 5.33. Let R be a simply connected, bounded region whose boundary B is a contour. Let $g(x,y)$ be defined and continuous on B. If $g_1(x,y), g_2(x,y), \ldots, g_n(x,y)$ are superfunctions of $g(x,y)$ on $R \cup B$, then

$$w(x,y) = \min [g_1(x,y), g_2(x,y), \ldots, g_n(x,y)], \qquad (x,y) \in (R \cup B) \tag{5.54}$$

is a superfunction of $g(x,y)$ on $R \cup B$.

Proof. By Theorem 5.28, $w(x,y)$ is superharmonic on $R \cup B$. The continuity of $g_1(x,y), g_2(x,y), \ldots, g_n(x,y)$ implies the continuity of $w(x,y)$. Finally, since $g_j(x,y) \geq g(x,y)$ on B, for $j = 1, 2, \ldots, n$, $w(x,y) \geq g(x,y)$ on B. Hence, $w(x,y)$, given by (5.54), is a superfunction of $g(x,y)$ on $R \cup B$ and the theorem is proved.

Theorem 5.34. Let R be a simply connected, bounded region whose boundary B is a contour. Let $g(x,y)$ be defined and continuous on B and let $w(x,y)$ be a superfunction of $g(x,y)$ on $R \cup B$. If R' is any circular region with boundary B' such that $R' \subset R$, then $M_{R'}[w(x,y)]$ is also a superfunction of $g(x,y)$ on $R \cup B$.

Proof. For convenience, set

$$M_{R'}[w(x,y)] = W(x,y).$$

Then, by definition, $W(x,y)$ is continuous on $R \cup B$. Since $W(x,y) = w(x,y)$ on B, $W(x,y) \geq g(x,y)$ on B. It remains to be shown only that $W(x,y)$ is superharmonic on $R \cup B$. For this purpose, select an arbitrary circular region R'' with boundary B'' such that $R'' \subset R$. It must be shown that on $R \cup B$

$$W(x,y) \geq M_{R''}[W(x,y)]. \tag{5.55}$$

For this purpose consider the following three cases.

CASE 1.   Let (x,y) not be an element of R''. Then $M_{R''}[W] = W$, so that (5.55) is valid.

CASE 2.   Let (x,y) be an element of R'' but not be an element of R'. Then

$$W(x,y) = M_{R'}[w(x,y)] = w(x,y) \geq M_{R''}[w(x,y)] \geq M_{R''}[W(x,y)],$$

so that (5.55) is again valid.

CASE 3.   Let $(x,y) \in (R' \cap R'')$. Then on the boundary of $R' \cap R''$ one has

$$W(x,y) \geq M_{R''}[W(x,y)], \tag{5.56}$$

since, on $(R' \cup B') \cap B''$, $W(x,y) = M_{R''}[W(x,y)]$, while on $(R'' \cup B'') \cap B'$,

$$M_{R''}[W(x,y)] = M_{R''}[w(x,y)] \leq w(x,y) = W(x,y).$$

Hence, on the boundary of $R' \cap R''$,

$$W(x,y) - M_{R''}[W(x,y)] \geq 0.$$

Moreover, on $R' \cap R''$, each of $W(x,y)$, $M_{R''}[W(x,y)]$ is harmonic, so that

$$W(x,y) - M_{R''}[W(x,y)]$$

is harmonic on $R' \cap R''$. Finally note that $W(x,y) - M_{R''}[W(x,y)]$ is continuous on $(R' \cup B') \cap (R'' \cup B'')$. Thus, by Theorem 5.3,

$$W(x,y) - M_{R''}[W(x,y)] \geq 0$$

on $R' \cap R''$, or, equivalently, that on $R' \cap R''$

$$W(x,y) \geq M_{R''}[W(x,y)].$$

Cases 1, 2, and 3 then imply that (5.55) is valid on $R \cup B$ and the theorem is proved.

We consider now a final supporting theorem.

Theorem 5.35.   Let R be a simply connected, bounded region whose boundary B is a contour. Let $(\bar{x},\bar{y}) \in B$. Then there exists a function $w(x,y)$ such that

(a) $w(x,y)$ is continuous on $R \cup B$
(b) $w(x,y)$ is superharmonic on $R \cup B$
(c) $w(\bar{x},\bar{y}) = 0$, and
(d) for $(x,y) \in (R \cup B)$, but different from $(\bar{x},\bar{y})$, $w(x,y) > 0$.

Proof. Let D be the diameter of the closed bounded set $R \cup B$, and without loss of generality, assume that $(\bar{x}, \bar{y})$ is the point $(0,0)$.

Let R_0 be the set $R \cup B$ with the point $(0,0)$ deleted. Then to each point $(x,y) \in R_0$ there corresponds a unique complex number $z = x + iy$. But to each such complex number $x + iy$, there corresponds an infinite number of angles $\phi + 2\pi k$; $k = 0, \pm 1, \pm 2, \ldots$, where ϕ is the polar angle of $x + iy$. Now, since $R \cup B$ is simply connected, we can and do select from the set of angles associated with each $(x,y) \in R_0$, a unique angle $\phi(x,y)$ such that $\phi(x,y)$ is continuous on R_0.

Then the function $w(x,y)$ defined by

$$w(x,y) = 0; \qquad (x,y) = (0,0), \tag{5.57}$$

$$w(x,y) = \frac{\log 2D - \log \sqrt{x^2 + y^2}}{(\log 2D - \log \sqrt{x^2 + y^2})^2 + [\phi(x,y)]^2},$$

$$(x,y) \neq (0,0), (x,y) \in (R \cup B), \tag{5.58}$$

is a function which satisfies the theorem. For, since $2D > \sqrt{x^2 + y^2}$, $(x,y) \in (R \cup B)$,

$$\log \frac{2D}{\sqrt{x^2 + y^2}} = \log 2D - \log \sqrt{x^2 + y^2} \neq 0,$$

so that the denominator of (5.58) is never zero. Moreover,

$$\lim_{(x,y) \to (0,0)} \frac{\log 2D - \log \sqrt{x^2 + y^2}}{(\log 2D - \log \sqrt{x^2 + y^2})^2 + [\phi(x,y)]^2} = 0,$$

so that (5.57) and (5.58) define a function which is continuous on $R \cup B$.

Direct calculation reveals that (5.58) is harmonic on R, so that $w(x,y)$, defined by (5.57) and (5.58) is superharmonic on $R \cup B$.

Finally note that for $(x,y) \neq (0,0)$, (5.58) implies $w(x,y) > 0$, and the theorem is proved.

The following major theorem will now establish the existence and uniqueness of the solution $u = f(x,y)$ of the Dirichlet problem.

Theorem 5.36. Let R be a simply connected, bounded region whose boundary B is a contour. Let $g(x,y)$ be defined and continuous on B. Then there exists on $R \cup B$ one and only one function $u = f(x,y)$ which is

(a) continuous on $R \cup B$,

(b) harmonic on R, and

(c) identical with $g(x,y)$ on B.

Proof. Let $f_\alpha(x,y)$ represent the set of all superfunctions of $g(x,y)$ on $R \cup B$ and, for $(x,y) \in (R \cup B)$, set

$$f(x,y) = \operatorname*{glb}_{\alpha} f_\alpha(x,y). \tag{5.59}$$

Recall that from Theorem 5.31 that if

$$c = \max_{B} |g(x,y)|, \tag{5.60}$$

then $-c \le f_\alpha(x,y)$ for all α, so that it is reasonable to discuss the greatest lower bound prescribed in (5.59).

We shall prove (*a*), (*b*), and (*c*) above, by establishing, in turn, that $f(x,y)$ is

(*a'*) continuous on R,
(*b'*) harmonic on R,
(*c'*) continuous on $R \cup B$ and identical with $g(x,y)$ on B.

Consider first (*a'*). Set $\bar{R} = R \cup B$. For fixed $d > 0$, let R_d be that subset of $\bar{R}$ such that $(x,y) \in R_d$ if and only if $(x,y) \in \bar{R}$ and the distance from (x,y) to B is greater than or equal to d. To show that $f(x,y)$ is continuous on R, it suffices to show that $f(x,y)$ is continuous on R_d for all positive d. Hence, let positive ϵ, d be given. It must be shown that if $(x_1,y_1) \in R_d$, then there exists δ such that if $(x_2,y_2) \in R_d$ and $(x_1 - x_2)^2 + (y_1 - y_2)^2 < \delta^2$, then $|f(x_1,y_1) - f(x_2,y_2)| < \epsilon$. Consider then an arbitrary point $(x_1,y_1) \in R_d$. Then there exists superfunction $f_\epsilon^*(x,y)$ of $g(x,y)$ on $R \cup B$ such that

$$f(x_1,y_1) \le f_\epsilon^*(x_1,y_1) < f(x_1,y_1) + \tfrac{1}{2}\epsilon. \tag{5.61}$$

Define

$$f_\epsilon(x,y) = \min [c, f_\epsilon^*(x,y)]. \tag{5.62}$$

Then, by Theorem 5.33, $f_\epsilon(x,y)$ is a superfunction of $g(x,y)$ on $R \cup B$ so that, by (5.59),

$$f(x_1,y_1) \le f_\epsilon(x_1,y_1). \tag{5.63}$$

Moreover, by (5.62), on $R \cup B$

$$f_\epsilon(x,y) \le f_\epsilon^*(x,y). \tag{5.64}$$

Thus, (5.61), (5.63), and (5.64) imply

$$f(x_1,y_1) \le f_\epsilon(x_1,y_1) < f(x_1,y_1) + \tfrac{1}{2}\epsilon. \tag{5.65}$$

Now let R' be a circular region with boundary B', center (x_1,y_1), radius d. Hence, $R' \subset R$. Define

$$f_\epsilon^{**}(x,y) = M_{R'}[f_\epsilon(x,y)]. \tag{5.66}$$

Then, by Theorem 5.34, $f_\epsilon^{**}(x,y)$ is a superfunction of $g(x,y)$ on $R \cup B$. But, by (5.59),

$$f(x_1,y_1) \leq f_\epsilon^{**}(x_1,y_1), \tag{5.67}$$

and by definition of superfunction

$$M_{R'}[f_\epsilon(x,y)] \leq f_\epsilon(x,y),$$

so that on $R \cup B$

$$f_\epsilon^{**}(x,y) \leq f_\epsilon(x,y). \tag{5.68}$$

Thus, (5.65), (5.67), and (5.68) imply

$$f(x_1,y_1) \leq f_\epsilon^{**}(x_1,y_1) \leq f_\epsilon(x_1,y_1) < f(x_1,y_1) + \tfrac{1}{2}\epsilon. \tag{5.69}$$

Since $f_\epsilon^{**}(x,y)$ is harmonic on R' and satisfies $|f_\epsilon^{**}(x,y)| \leq c$ on $R \cup B$, it follows from Theorem 5.18 that on the circular region R'' with boundary B'', radius $\tfrac{1}{2}d$, center (x_1,y_1), one can find an upper bound which depends only on c and d of the first-order partial derivatives of $f_\epsilon^{**}(x,y)$. Hence, there exists positive $\delta < \tfrac{1}{2}d$ such that if (x_2,y_2) is any point such that $(x_1 - x_2)^2 + (y_1 - y_2)^2 < \delta^2$, then

$$|f_\epsilon^{**}(x_1,y_1) - f_\epsilon^{**}(x_2,y_2)| < \tfrac{1}{2}\epsilon. \tag{5.70}$$

Moreover, the number δ depends *only* on c, d, and ϵ (consult Courant [7], vol. 2, pp. 54–55). Therefore, by (5.59), (5.69), and (5.70),

$$f(x_2,y_2) \leq f_\epsilon^{**}(x_2,y_2) < f_\epsilon^{**}(x_1,y_1) + \tfrac{1}{2}\epsilon < f(x_1,y_1) + \epsilon. \tag{5.71}$$

By a symmetrical argument, it follows that, if $(x_2,y_2) \in R_d$ and $(x_2 - x_1)^2 + (y_2 - y_1)^2 < \delta^2$, one has

$$f(x_1,y_1) \leq f(x_2,y_2) + \epsilon. \tag{5.72}$$

Thus, for any two points (x_1,y_1), (x_2,y_2) of R_d which satisfy $(x_2 - x_1)^2 + (y_2 - y_1)^2 < \delta^2$, one has by (5.71) and (5.72) that

$$|f(x_1,y_1) - f(x_2,y_2)| < \epsilon,$$

from which the assertion of continuity on R_d, and hence on R, follows.

Consider next (b'). We shall show $f(x,y)$ is harmonic on R by showing that if R' is any circular region with boundary B' such that $(R' \cup B') \subset R$, then $f(x,y)$ is harmonic on R'. Given $\epsilon > 0$ and $(x_1,y_1) \in (R' \cup B')$, there exists superfunction $f_\epsilon(x,y)$ of $g(x,y)$ on $R \cup B$ such that

$$f(x_1,y_1) \leq f_\epsilon(x_1,y_1) < f(x_1,y_1) + \epsilon. \tag{5.73}$$

Moreover, by (5.59) and since $f(x,y)$, $f_\epsilon(x,y)$ are continuous on R, there exists circular region R'' with boundary B'', center (x_1,y_1), such that on R''

$$f(x,y) \le f_\epsilon(x,y) < f(x,y) + \epsilon. \tag{5.74}$$

In general, then, for each $(x_\alpha,y_\alpha) \in (R' \cup B')$, there exists superfunction $f_{\epsilon,\alpha}(x,y)$, circular region R_α with boundary B_α, center (x_α,y_α) such that on R_α

$$f(x,y) \le f_{\epsilon,\alpha}(x,y) < f(x,y) + \epsilon. \tag{5.75}$$

Since $\underset{\alpha}{\cup} R_\alpha$ is an open covering of the closed, bounded set $R' \cup B'$, the Heine-Borel theorem implies there exists a finite number of the R_α, say, R_{α_1}, R_{α_2}, ..., R_{α_n} which cover $R' \cup B'$.

Define

$$f_\epsilon^*(x,y) = \min [f_{\epsilon,\alpha_1}(x,y), f_{\epsilon,\alpha_2}(x,y), \ldots, f_{\epsilon,\alpha_n}(x,y)].$$

By Theorem 5.33, $f_\epsilon^*(x,y)$ is a superfunction of $g(x,y)$ on $R \cup B$ and by construction

$$f(x,y) \le f_\epsilon^*(x,y) < f(x,y) + \epsilon.$$

Define now

$$f_\epsilon^{**}(x,y) = M_{R'}[f_\epsilon^*(x,y)],$$

so that on $R' \cup B'$

$$f(x,y) \le f_\epsilon^{**}(x,y) < f(x,y) + \epsilon. \tag{5.76}$$

From (5.76), then, $f_\epsilon^{**}(x,y)$ converges uniformly to $f(x,y)$ as $\epsilon \to 0$. But $f_\epsilon^{**}(x,y)$ is harmonic on R', so that, by Harnack's theorem, $f(x,y)$ is harmonic on R', and hence on R.

Finally, consider (c'). Let $\epsilon > 0$ and let $(x_\beta,y_\beta) \in B$. Define positive γ so large that for $(x,y) \in B$,

$$\epsilon + \gamma w(x,y) > |g(x,y) - g(x_\beta,y_\beta)|,$$

where $w(x,y)$ is given by Theorem 5.35 so that $w(x_\beta,y_\beta) = 0$. Define further

$$f_1(x,y) = g(x_\beta,y_\beta) + \gamma w(x,y) + \epsilon. \tag{5.77}$$

Then $f_1(x,y)$ is a superfunction of $g(x,y)$ on $R \cup B$. Moreover,

$$f_2(x,y) = g(x_\beta,y_\beta) - \gamma w(x,y) - \epsilon \tag{5.78}$$

is a subfunction of $g(x,y)$ on $R \cup B$. Hence, on $R \cup B$,

$$f_2(x,y) \le f(x,y) \le f_1(x,y). \tag{5.79}$$

But, $w(x,y)$ is continuous on $R \cup B$, so that for the given $\epsilon > 0$, there exists δ such that for all $(x,y) \in (R \cup B)$ such that

$$(x - x_\beta)^2 + (y - y_\beta)^2 < \delta^2,$$

then $\gamma w(x,y) < \epsilon$.   Hence, for such values of (x,y), (5.77) to (5.79) imply

$$g(x_\beta,y_\beta) - 2\epsilon \le f(x,y) \le g(x_\beta,y_\beta) + 2\epsilon.$$

By taking the limit as $\epsilon \to 0$ of each term of the latter inequality, it follows that since (x_β,y_β) may be selected arbitrarily on B, that $f(x,y)$ is continuous on $R \cup B$ and that $f(x,y) \equiv g(x,y)$ on B.   Thus (c') is valid.

Since the uniqueness of the solution of the Dirichlet problem was established in Theorem 5.6, the theorem is proved.

5.5 Dirichlet Problems on a Square

An even more difficult problem than that of establishing the *existence* of the unique solution of a given Dirichlet problem is that of *producing* the solution itself.   If R is a circular region, then the solution can be given, as shown in Sec. 5.3, by means of an integral or by means of an equivalent Fourier series.   We now show, by examples, how to solve by means of the method of separation of variables a special class of Dirichlet problems for which R is a square region.

Example 1. Let B be the square with vertices $(0,0)$, $(\pi,0)$, (π,π), $(0,\pi)$ and let R be its interior.   Let $g(x) = \sin x$, $0 \le x \le \pi$.   Then find the unique function $u = f(x,y)$ which is continuous on $R \cup B$ and satisfies

$$u_{xx} + u_{yy} = 0, \qquad \text{on } R \tag{5.80}$$

$$f(x,0) = g(x), \qquad 0 \le x \le \pi \tag{5.81}$$

$$f(\pi,y) = 0, \qquad 0 \le y \le \pi \tag{5.82}$$

$$f(x,\pi) = 0, \qquad 0 \le x \le \pi \tag{5.83}$$

$$f(0,y) = 0, \qquad 0 \le y \le \pi. \tag{5.84}$$

Solution.   Assume a solution of the problem of the form

$$f(x,y) = X(x)Y(y). \tag{5.85}$$

Then (5.85) *cannot be identically zero* if it is to satisfy (5.81).   Substitution of (5.85) into (5.80) and separating the variables yields

$$\frac{d^2X}{dx^2} + \lambda X = 0, \tag{5.86}$$

$$\frac{d^2Y}{dy^2} - \lambda Y = 0, \tag{5.87}$$

where λ is an arbitrary constant. If one sets $X(0) = X(\pi) = 0$ so as to satisfy (5.82) and (5.84), then, in a fashion analogous to that described for (4.51), (5.86) implies that λ must be positive. The general solution of (5.86) is then

$$X(x) = A \sin \sqrt{\lambda} x + B \cos \sqrt{\lambda} x, \qquad (5.88)$$

where A, B are arbitrary constants. The condition $X(0) = 0$ implies with respect to (5.88) that $B = 0$. Since one does not wish $X(x) \equiv 0$, A cannot be zero. Thus, to satisfy $X(\pi) = 0$, the implication that λ must be positive and (5.88) imply that one must select $\lambda = n^2, n = 1, 2, \ldots$. If $X_n(x)$ is the solution $X(x)$ of (5.86) which corresponds to $\lambda = n^2$, $n = 1, 2, \ldots$, one has finally

$$X_n(x) = A_n \sin nx, \qquad n = 1, 2, \ldots, \qquad (5.89)$$

where for each value of n, A_n is an arbitrary constant.

For $\lambda = n^2, n = 1, 2, 3, \ldots$, the general solution of (5.87) corresponding to each n is

$$Y_n(y) = C_n e^{ny} + D_n e^{-ny}.$$

For simplicity, set $C_n = \frac{1}{2} e^{c_n n}$, $D_n = -\frac{1}{2} e^{-c_n n}$, where c_n is arbitrary so that

$$Y_n(y) = \tfrac{1}{2}(e^{ny + c_n n} - e^{-ny - c_n n}) = \sinh(ny + c_n n). \qquad (5.90)$$

Thus, (5.89) and (5.90) imply that for $n = 1, 2, \ldots$

$$f_n(x,y) = X_n(x) Y_n(y) = A_n \sin nx \sinh(ny + c_n n) \qquad (5.91)$$

are harmonic functions which satisfy (5.82) and (5.84).

If each function of (5.91) were to satisfy (5.83), then $c_n = -\pi$, $n = 1, 2, \ldots$, so that (5.91) becomes

$$f_n(x,y) = A_n \sin nx \sinh(ny - n\pi). \qquad (5.92)$$

Consider now an infinite series of harmonic functions of the form

$$f(x,y) = \sum_{n=1}^{\infty} A_n \sin nx \sinh(ny - n\pi). \qquad (5.93)$$

Assuming the convergence of (5.93), one would have $f(0,y) = f(\pi,y) = f(x,\pi) = 0$, thus satisfying (5.82) to (5.84). In order to satisfy (5.81), it is necessary that

$$f(x,0) \equiv \sum_{n=1}^{\infty} A_n \sin nx \sinh(-\pi n) \equiv \sin x,$$

so that $A_1 = -(\sinh \pi)^{-1}$; $A_{n+1} = 0, n = 1, 2, 3, \ldots$. Thus (5.93) takes the form

$$f(x,y) = -\frac{\sin x \sinh(y - \pi)}{\sinh \pi} \qquad (5.93')$$

and presents itself as a solution of the problem.   Direct calculation reveals
that (5.93') satisfies (5.80) to (5.84) and is therefore the unique solution
of the problem.

Example 2.   Let B be the square with vertices $(0,0)$, $(\pi,0)$, (π,π), $(0,\pi)$
and let R be its interior.   Let $g(x) = 1 - \cos 2x$, $0 \le x \le \pi$.   Then find
the unique solution $u = f(x,y)$ which is continuous on $R \cup B$ and satisfies
(5.80) to (5.84).

Solution.   Since $g(x)$ does not enter the discussion of Example 1 until
after the formulation of (5.93), it follows that the discussion for Example 2
is identical to that for Example 1 up to and including (5.93).   In order then
to have (5.93) satisfy (5.81), it is necessary that

$$f(x,0) \equiv \sum_{n=1}^{\infty} A_n \sin nx \sinh (-\pi n) \equiv 1 - \cos 2x, \qquad 0 \le x \le \pi,$$

or, equivalently, that $-A_n \sinh n\pi$, $n = 1, 2, \ldots$, must be the Fourier
sine coefficients of the function $1 - \cos 2x$.   The Fourier sine coefficients
of $1 - \cos 2x$ were given previously [see (4.72)] and are, for completeness,
restated below.   If

$$1 - \cos 2x \equiv \sum_{k=1}^{\infty} b_k^* \sin kx, \qquad 0 \le x \le \pi, \tag{5.94}$$

then

$$b_1^* = \frac{16}{3\pi}, \qquad b_2^* = 0,$$

$$\left. \begin{array}{l} b_k^* = \frac{2}{\pi}\left[\frac{1}{k} - \frac{\cos k\pi}{k} + \frac{\cos (k-2)\pi}{2(k-2)} + \frac{\cos (k+2)\pi}{2(k+2)} - \frac{1}{2(k-2)} \right. \\ \\ \left. \qquad - \frac{1}{2(k+2)}\right], \qquad k = 3, 4, 5, \ldots, \end{array} \right\} \tag{5.95}$$

so that

$$A_n = \frac{b_n^*}{-\sinh n\pi}, \qquad n = 1, 2, 3, \ldots.$$

Thus, (5.93) presents itself as a solution of the problem under consideration
in the form

$$f(x,y) = \sum_{n=1}^{\infty} \left[b_n^* \sin nx \, \frac{\sinh (n\pi - ny)}{\sinh n\pi} \right], \tag{5.96}$$

where b_n^*, $n = 1, 2, \ldots$, are given by (5.95).   We must now show that
(5.96) is the desired solution.

By Corollary 2 to Theorem 2.6, the series

$$\sum_{k=1}^{\infty} b_k^* \sin kx$$

converges absolutely and uniformly to $1 - \cos 2x$ on $0 \le x \le \pi$. By Theorem 2.5, it follows from (5.94) that

$$\frac{d}{dx}(1 - \cos 2x) \equiv \sum_{k=1}^{\infty} kb_k^* \cos kx, \qquad 0 < x < \pi.$$

By Corollary 1 to Theorem 2.6, it follows for the given example that

$$\sum_{k=1}^{\infty} |kb_k^*|$$

converges.   Hence, by the Weierstrass M-test, the series

$$\sum_{k=1}^{\infty} kb_k^* \cos kx$$

converges absolutely and uniformly for all real x.   Since $\dfrac{d}{dx}(1 - \cos 2x)$ is continuous on $0 \le x \le \pi$, it follows that

$$\frac{d}{dx}(1 - \cos 2x) \equiv \sum_{k=1}^{\infty} kb_k^* \cos kx, \qquad 0 \le x \le \pi, \qquad (5.97)$$

and that the convergence is absolute and uniform.
    Similarly, by the Weierstrass M-test, the series

$$\sum_{k=1}^{\infty} kb_k^* \sin kx \qquad\qquad (5.98)$$

converges absolutely and uniformly on $0 \le x \le \pi$.
    With the above results available, we shall attempt to establish the uniform convergence of (5.96) by applying Abel's test.   For this purpose, consider first the sequence of functions

$$w_n(y) = \frac{\sinh(n\pi - ny)}{\sinh n\pi}, \qquad 0 \le y \le \pi; \qquad n = 1, 2, 3, \dots, \qquad (5.99)$$

and let us show first that $w_n(y)$ is *monotonic decreasing with n* for all y in $0 \le y \le \pi$.
    Since (5.99) is continuous for each n on $0 \le y \le \pi$, it is only necessary to show that $w_n(y)$ is monotonic decreasing with n for all y in $0 < y < \pi$ in order to establish the described monotonicity on $0 \le y \le \pi$.   Since $0 < y < \pi$, then $0 < \pi - y < \pi$, so that $w_n(y)$ would be monotonic decreasing with n on $0 < y < \pi$ provided the function

$$W(t) = \frac{\sinh \alpha t}{\sinh \beta t}, \qquad 0 < \alpha < \beta, t > 0,$$

is a decreasing function of t.   However,

$$\frac{dW}{dt} = \frac{\alpha \cosh \alpha t \sinh \beta t - \beta \sinh \alpha t \cosh \beta t}{\sinh^2 \beta t}, \qquad t > 0.$$

Set $\alpha = \beta - \delta$, $\delta > 0$, so that

$$\frac{dW}{dt} = \frac{\beta \sinh(\alpha t - \beta t) - \delta \cosh \alpha t \sinh \beta t}{\sinh^2 \beta t}, \qquad t > 0.$$

Since $\alpha t - \beta t < 0$, $\sinh(\alpha t - \beta t) < 0$, so that $\beta \sinh(\alpha t - \beta t) < 0$ and $-\delta \cosh \alpha t \sinh \beta t < 0$. Hence, $dW/dt < 0$ so that for $t > 0$, $W(t)$ is a decreasing function of t and (5.99) is monotonic decreasing with n for all y in $0 \le y \le \pi$.

Similarly,

$$v_n(y) = \frac{\cosh(n\pi - ny)}{\sinh n\pi}, \qquad 0 \le y \le \pi, n = 1, 2, \ldots, \quad (5.100)$$

is also monotonic decreasing with n for $0 \le y \le \pi$.

Now, the functions $w_n(y)$ are all nonnegative and less than or equal to unity for all n and all y under consideration, so that they are uniformly bounded. Similarly, the functions $v_n(y)$ are uniformly bounded. Hence, by the uniform convergence of (5.94), (5.97), and (5.98), Abel's test implies the uniform convergence in x and y together on $R \cup B$ of (5.96) and of the series

$$f_x(x,y) \equiv \sum_{n=1}^{\infty} \left[nb_n^* \cos nx \frac{\sinh(n\pi - ny)}{\sinh n\pi} \right], \tag{5.101}$$

$$f_y(x,y) \equiv \sum_{n=1}^{\infty} \left[-nb_n^* \sin nx \frac{\cosh(n\pi - ny)}{\sinh n\pi} \right]. \tag{5.102}$$

In a similar fashion it follows that on R

$$f_{xx}(x,y) \equiv - \sum_{n=1}^{\infty} \left[n^2 b_n^* \sin nx \frac{\sinh(n\pi - ny)}{\sinh n\pi} \right], \tag{5.103}$$

$$f_{yy}(x,y) \equiv \sum_{n=1}^{\infty} \left[n^2 b_n^* \sin nx \frac{\sinh(n\pi - ny)}{\sinh n\pi} \right]. \tag{5.104}$$

Hence, (5.103) and (5.104) imply that $f(x,y)$ is harmonic on R, while (5.96) is readily shown to satisfy (5.81) to (5.84), where $g(x)$ is defined by (5.94). Moreover, by the uniformity of the convergence of (5.96), $f(x,y)$ is continuous on $R \cup B$. Thus (5.96) is the unique solution of the Dirichlet problem under consideration.

EXERCISES

5.1. Show that if (r,ϕ) are polar coordinates of (x,y) then, except at the point $(x,y) = (0,0)$, the potential equation is equivalent to

$$r^2 \frac{\partial^2 u}{\partial r^2} + r \frac{\partial u}{\partial r} + \frac{\partial^2 u}{\partial \phi^2} = 0.$$

5.2. Let R be a simply connected, bounded region whose boundary B is a contour. Let $u = f(x,y)$ be harmonic on R and continuous on $\bar{R} = R \cup B$. Show that if there exists a point $(\bar{x}, \bar{y}) \in R$ such that $f(\bar{x}, \bar{y}) = \underset{R}{\text{lub}}\ f(x,y)$, then on $R \cup B$, $f(x,y)$ is identically constant.

5.3. Show that if $u = f(x,y)$ is harmonic on E^2, then on E^2 the function under consideration is not bounded above or below.

5.4. Let R be the circular region with boundary B, center $(0,0)$, and radius unity. On $R \cup B$ find in terms of the Poisson integral the unique solution of the Dirichlet problem which corresponds to each of the following functions $g(x,y)$ defined on B:

(a) $g(x,y) = 1 + x$ $\left(Ans.\ \dfrac{1}{2\pi} \displaystyle\int_{-\pi}^{\pi} \dfrac{(1 + \cos t)(1 - x^2 - y^2)\, dt}{1 - 2x \cos t - 2y \sin t + x^2 + y^2} \right)$

(b) $g(x,y) = x + y^2$ $\left(Ans.\ \dfrac{1}{2\pi} \displaystyle\int_{-\pi}^{\pi} \dfrac{(\cos t + \sin^2 t)(1 - x^2 - y^2)\, dt}{1 - 2x \cos t - 2y \sin t + x^2 + y^2} \right)$

(c) $g(x,y) = \sin y$

(d) $g(x,y) = 1$

(e) $g(x,y) = x^2 - y^2$

(f) $g(x,y) = x^2 y^3$

(g) $g(x,y) = \dfrac{xy}{1 + y^2}$.

5.5. Find in terms of Fourier series the unique solution of each of the Dirichlet problems described in Exercise 5.4, above.

5.6. Let R be the circular region with boundary B, center $(0,0)$, and radius r. On $R \cup B$ find in terms of an integral the unique solution of the Dirichlet problem which corresponds to each of the following functions $g(x,y)$ defined on B and to each indicated radius r:

(a) $g(x,y) = x$, $\qquad r = 2$

$\left(Ans.\ \dfrac{1}{2\pi} \displaystyle\int_{-\pi}^{\pi} \dfrac{2 \cos t\, (4 - x^2 - y^2)\, dt}{4 - 4x \cos t - 4y \sin t + x^2 + y^2} \right)$

(b) $g(x,y) = x^2 y$, $\qquad r = 3$

(c) $g(x,y) = \sin y$, $\qquad r = 4$

(d) $g(x,y) = 1 + x - y^2$, $\quad r = 5$.

5.7. Find in terms of Fourier series the unique solution of each of the Dirichlet problems described in Exercise 5.6, above.

5.8. Let R be the circular region with boundary B, center $(-2,1)$, and radius r. On $R \cup B$ find in terms of an integral the unique solution of the Dirichlet problem which corresponds to each of the following functions $g(x,y)$ defined on B and each indicated radius r:

(a) $g(x,y) = x$, $\qquad r = 2$

(b) $g(x,y) = x^2 y$, $\qquad r = 3$

(c) $g(x,y) = \sin y$, $\qquad r = 4$

(d) $g(x,y) = 1 + x - y^2$, $\quad r = 5$.

5.9. Let R be a simply connected, bounded region whose boundary B is a contour. Show that if $u = f(x,y)$ is continuous on $R \cup B$ and harmonic on R, then, on R, $f(x,y)$ has all partial derivatives of all orders, which themselves are harmonic on R.

5.10. Let R be a simply connected, bounded region whose boundary B is a contour. Let $u = f(x,y)$ be harmonic on R and continuous on $R \cup B$. By the continuity property, there exists positive number M such that for $(x,y) \in (R \cup B)$, $|f(x,y)| < M$. Then show that on every closed subset $R^* \subset R$, there exists a uniform bound for the nth-order partial derivatives of $f(x,y)$ of the form

$$\left| \frac{\partial^n f(x,y)}{\partial x^i \, \partial y^j} \right| \le M \left(\frac{4}{\pi \rho} \right)^n n^n, \qquad i + j = n,$$

where ρ is the positive distance between the two closed, bounded sets R^* and B.

5.11. Let B be the unit circle and let R be its interior. On $R \cup B$ find, if possible, an example of a subharmonic function $f_1(x,y)$ and an example of a superharmonic function $f_2(x,y)$, neither of which is harmonic on R.

5.12. Let R be a simply connected, bounded region whose boundary B is a contour. On $R \cup B$, let $v(x,y)$ be superharmonic and $w(x,y)$ be subharmonic. Show that $v(x,y) - w(x,y)$ is superharmonic on $R \cup B$.

5.13. Carry out the details in showing that (5.58) is harmonic for all $(x,y) \ne (0,0)$. (*Hint:* Use polar coordinates.)

5.14. Discuss the possibility of extending Theorem 5.36 to a region R whose boundary B is not necessarily a contour.

5.15. Let B be the square with vertices $(0,0)$, $(\pi,0)$, (π,π), $(0,\pi)$ and let R be its interior. Consider the Dirichlet problem of finding a function $u = f(x,y)$ which is harmonic on R and satisfies

$$\begin{aligned}
f(x,0) &= g(x), & 0 &\le x \le \pi, \\
f(\pi,y) &= 0, & 0 &\le y \le \pi, \\
f(x,\pi) &= 0, & 0 &\le x \le \pi, \\
f(0,y) &= 0, & 0 &\le y \le \pi.
\end{aligned}$$

Then for each of the following functions $g(x)$ find the corresponding $u = f(x,y)$ and check your result:

(a) $g(x) = \sin 2x$

(b) $g(x) = \sin x + 2 \sin 2x$

(c) $g(x) = \sin x + 3 \sin 2x + \sin 3x$

(d) $g(x) = \sum\limits_{k=1}^{n} \sin kx;$ n a fixed positive integer

(e) $g(x) = \sin^2 x$

(f) $g(x) = 1 - \cos 4x$

(g) $g(x) = -\pi^2 x^2 + 2\pi x^3 - x^4.$

5.16. Let B be the square with vertices $(0,0)$, $(\pi,0)$, (π,π), $(0,\pi)$, and let R be its interior. Consider the Dirichlet problem of finding a function $u = f(x,y)$ which is harmonic on R and satisfies

$$\begin{aligned}
f(x,0) &= g_1(x), & 0 &\le x \le \pi, \\
f(\pi,y) &= g_2(y), & 0 &\le y \le \pi, \\
f(x,\pi) &= g_3(x), & 0 &\le x \le \pi, \\
f(0,y) &= g_4(y), & 0 &\le y \le \pi.
\end{aligned}$$

Then for each of the following sets of functions $g_1(x), g_2(y), g_3(x), g_4(y)$, find the corresponding $u = f(x,y)$ and check your result:

(a) $g_1(x) = \sin x,$ $\quad g_2(y) = 0,$ $\qquad g_3(x) = 0,$ $\qquad g_4(y) = 0$

(b) $g_1(x) = 0,$ $\qquad g_2(y) = \sin y,$ $\qquad g_3(x) = 0,$ $\qquad g_4(y) = 0$

(c) $g_1(x) = \sin x,$ $\quad g_2(y) = \sin y,$ $\qquad g_3(x) = 0,$ $\qquad g_4(y) = 0$

(d) $g_1(x) = 0,$ $\qquad g_2(y) = 0,$ $\qquad\quad g_3(x) = \sin 2x,$ $\quad g_4(y) = 0$

(e) $g_1(x) = \sin x,$ $\quad g_2(y) = \sin y,$ $\qquad g_3(x) = \sin 2x,$ $\quad g_4(y) = 0$

(f) $g_1(x) = 0,$ $\qquad g_2(y) = 0,$ $\qquad\quad g_3(x) = 0,$ $\qquad g_4(y) = \sin 3y$

(g) $g_1(x) = \sin x,$ $\quad g_2(y) = \sin y,$ $\qquad g_3(x) = \sin 2x,$ $\quad g_4(y) = \sin 3y$

(h) $g_1(x) = \sin x,$ $\quad g_2(y) = 1 - \cos^2 y,$ $\quad g_3(x) = \sin^2 x,$ $\quad g_4(y) = 1 - \cos 4y.$

5.17. Let B be the rectangle with vertices $(0,0), (a,0), (a,b), (0,b); a > 0, b > 0$, and let R be its interior. Consider the Dirichlet problem of finding a function $u = f(x,y)$ which is harmonic on R and satisfies

$$f(x,0) = g_1(x), \qquad 0 \leq x \leq a,$$
$$f(a,y) = g_2(y), \qquad 0 \leq y \leq b,$$
$$f(x,b) = g_3(x), \qquad 0 \leq x \leq a,$$
$$f(0,y) = g_4(y), \qquad 0 \leq y \leq b.$$

Then for each of the following sets of functions $g_1(x), g_2(y), g_3(x), g_4(y)$ and the given values of a, b, find the corresponding $u = f(x,y)$, and check your result:

(a) $g_1(x) = 2 \sin x,$ $\qquad g_2(y) = 0,$ $\qquad g_3(x) = 0,$
$\qquad\qquad\qquad\qquad\qquad\qquad\qquad g_4(y) = 0;$ $\qquad a = \pi, b = 2$

(b) $g_1(x) = \sin 2x,$ $\qquad g_2(y) = 0,$ $\qquad g_3(x) = 0,$
$\qquad\qquad\qquad\qquad\qquad\qquad\qquad g_4(y) = 0;$ $\qquad a = \pi, b = 4$

(c) $g_1(x) = 1 - \cos 2x,$ $\quad g_2(y) = 0,$ $\qquad g_3(x) = 0,$
$\qquad\qquad\qquad\qquad\qquad\qquad\qquad g_4(y) = 0;$ $\qquad a = \pi, b = 1$

(d) $g_1(x) = 0,$ $\qquad\qquad g_2(y) = \sin y,$ $\quad g_3(x) = 0,$
$\qquad\qquad\qquad\qquad\qquad\qquad\qquad g_4(y) = 0;$ $\qquad a = 2, b = \pi$

(e) $g_1(x) = \sin x,$ $\qquad g_2(y) = 0,$ $\qquad g_3(x) = \sin 2x,$
$\qquad\qquad\qquad\qquad\qquad\qquad\qquad g_4(y) = 0;$ $\qquad a = \pi, b = 2$

(f) $g_1(x) = \sin 2\pi x,$ $\quad g_2(y) = 0,$ $\qquad g_3(x) = 0,$
$\qquad\qquad\qquad\qquad\qquad\qquad\qquad g_4(y) = 0;$ $\qquad a = 1, b = 2$

(g) $g_1(x) = \sin 2\pi x,$ $\quad g_2(y) = \sin 2\pi y,$ $\quad g_3(x) = 0,$
$\qquad\qquad\qquad\qquad\qquad\qquad\qquad g_4(y) = 0;$ $\qquad a = 1, b = 1$

(h) $g_1(x) = \sin 2\pi x,$ $\quad g_2(y) = \sin 2\pi y,$ $\quad g_3(x) = \sin 4\pi x,$
$\qquad\qquad\qquad\qquad\qquad\qquad\qquad g_4(y) = \sin 4\pi y;$ $\quad a = 1, b = 1.$

5.18. Prove Theorem 5.12 directly by means of Theorem 5.2.

5.19. Let R be a simply connected, bounded region whose boundary B is a contour. Let $g(x,y)$ be defined and continuous on B. The problem of finding a function $u = f(x,y)$ which is harmonic on $E^2 - (R \cup B)$, continuous on $E^2 - R$, and identical with $g(x,y)$ on B is called an *exterior* Dirichlet problem. Show that the exterior Dirichlet problem has a unique solution $u = f(x,y)$ in the class of functions which are bounded on $E^2 - R$.

5.20. Let R be a simply connected, bounded region whose boundary B is smooth. Let $g(x,y)$ be defined and continuous on B. The problem of finding a function $u = f(x,y)$ which is harmonic on R and continuous on $R \cup B$, and whose normal derivative $\partial f/\partial n$ in the direction of the inner normal is identical with $g(x,y)$ at every point of B is called a Neumann problem.

(a) Show that for the Neumann problem to have a solution it is necessary that

$$\oint_B g(x,y)\, ds = 0.$$

(b) Show that if the Neumann problem has a solution, then it has an infinite number of solutions, any two of which differ only by an additive constant.

5.21. Let R be a simply connected, bounded region whose boundary B is a contour. Let $g(x,y)$ be defined and continuous on B and let $G(x,y)$ be continuous on $R \cup B$ and of class C^1 on R. Show that on $R \cup B$ there exists a unique function $u = f(x,y)$ which is continuous on $R \cup B$, satisfies the Poisson equation

$$\frac{\partial^2 u}{\partial x^2} + \frac{\partial^2 u}{\partial y^2} = G(x,y)$$

on R, and is identical with $g(x,y)$ on B.

5.22. Let R be a simply connected, bounded region whose boundary B is a contour. Let $u = f(x,y)$ be continuous on $R \cup B$ and harmonic on R. Show that for each circular region R_1 with boundary B_1, radius r_1, and center (x_1, y_1) such that $(R_1 \cup B_1) \subset (R \cup B)$, it is necessary that

$$f(x_1, y_1) = \frac{1}{2\pi r_1} \oint_{B_1} f(x,y)\, ds.$$

5.23. Let R be the set of points (x,y) whose coordinates satisfy $x^2 + y^2 < 1$ and let B be the boundary of R. Let R_1 be the set of points (x,y) whose coordinates satisfy $0 < x^2 + y^2 < 1$. Then show that there does not exist a function $u = f(x,y)$ which is continuous on $R \cup B$, assumes the value 1 at $(0,0)$, assumes the value 0 everywhere on B, and is harmonic on R_1. (Note that this example indicates the impossibility of extending Theorem 5.36 to *all connected* regions. Theorem 5.36 can, however, be extended to a large class of regions which are of importance in the physical sciences and which are *not* simply connected. Consult, for example, Petrovsky [37], p. 187.)

5.24. Show that all the derivatives of a harmonic function are also harmonic functions.

Chapter 6

THE HEAT EQUATION

Problems analogous to those considered for equations of elliptic type are also of interest for equations of parabolic type, and those which we consider below are intimately associated with the physics of heat conduction.

6.1 The Boundary-value Problem of Type H

Definition 6.1. The parabolic partial differential equation

$$\frac{\partial^2 u}{\partial x^2} = \frac{\partial u}{\partial y}$$

is called the heat equation.

For $b > 0$, let B be the rectangle with vertices $(0,0)$, $(\pi,0)$, (π,b), $(0,b)$, and let R be its interior. Denote the straight-line segments connecting $(0,b)$ and $(0,0)$, $(0,0)$ and $(\pi,0)$, $(\pi,0)$ and (π,b), (π,b) and $(0,b)$ by L_1, L_2, L_3, L_4, respectively, as shown in Diagram 6.1. Set

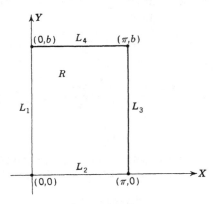

DIAGRAM 6.1

L_4 minus its end points $(0,b)$, (π,b) equal to L_4^*. Let $g_1(y)$, $0 \le y \le b$; $g_2(x)$, $0 \le x \le \pi$; $g_3(y)$, $0 \le y \le b$ be defined and continuous and let these functions satisfy $g_1(0) = g_2(0)$, $g_2(\pi) = g_3(0)$. Then the boundary-value problem of type H requires that one find a function $u = f(x,y)$ which

(a) is continuous on $R \cup B$,

(b) satisfies the heat equation

$$u_{xx} = u_y \tag{6.1}$$

on $R \cup L_4^*$, and

(c) satisfies

$$f(0,y) \equiv g_1(y) \text{ on } L_1, \tag{6.2}$$

$$f(x,0) \equiv g_2(x) \text{ on } L_2, \tag{6.3}$$

$$f(\pi,y) \equiv g_3(y) \text{ on } L_3. \tag{6.4}$$

Theorem 6.1.   For $b > 0$, let B be the rectangle with vertices $(0,0)$, $(\pi,0)$, (π,b), $(0,b)$, and let R be its interior.   If $u = f(x,y)$ is continuous on $R \cup B$ and is a solution of (6.1) on $R \cup L_4^*$, then $f(x,y)$ takes on its maximum value on $L_1 \cup L_2 \cup L_3$.

Proof.   Since $f(x,y)$ is continuous on $R \cup B$, set

$$\max_{R \cup B} f(x,y) = M, \tag{6.5}$$

$$\max_{L_1 \cup L_2 \cup L_3} f(x,y) = m. \tag{6.6}$$

Assume that

$$M > m. \tag{6.7}$$

Then, there exists a point $(x_1,y_1) \in (R \cup L_4^*)$ at which $f(x_1,y_1) = M$. On $R \cup B$ define the function

$$v(x,y) = f(x,y) + \frac{M - m}{4\pi^2}(x - x_1)^2. \tag{6.8}$$

On $L_1 \cup L_2 \cup L_3$, it follows that

$$v(x,y) \le m + \frac{M - m}{4} = \frac{M}{4} + \frac{3m}{4} = \alpha M, \qquad 0 < \alpha < 1. \tag{6.9}$$

Also,

$$v(x_1,y_1) = M. \tag{6.10}$$

Thus $v(x,y)$ also assumes its maximum value V on $R \cup L_4^*$.   Suppose then at $(x_2,y_2) \in (R \cup L_4^*)$, $v(x_2,y_2) = V$.   Then, if $(x_2,y_2) \in R$

$$\frac{\partial v(x_2,y_2)}{\partial y} = 0,$$

while if $(x_2, y_2) \in L_4^*$,

$$\frac{\partial v(x_2, y_2)}{\partial y} \geq 0,$$

so that if $(x_2, y_2) \in (R \cup L_4^*)$,

$$\frac{\partial v(x_2, y_2)}{\partial y} \geq 0. \tag{6.11}$$

From (6.8), then,

$$\frac{\partial f}{\partial y} = \frac{\partial v}{\partial y} \tag{6.12}$$

and

$$\frac{\partial^2 v}{\partial x^2} = \frac{\partial^2 f}{\partial x^2} + \frac{M - m}{2\pi^2}. \tag{6.13}$$

Now (6.11) and (6.12) imply

$$\frac{\partial f(x_2, y_2)}{\partial y} \geq 0, \tag{6.14}$$

while (6.14) and (6.1) imply

$$\frac{\partial^2 f(x_2, y_2)}{\partial x^2} \geq 0. \tag{6.15}$$

Finally, (6.13) and (6.15) imply

$$\frac{\partial^2 v(x_2, y_2)}{\partial x^2} > 0. \tag{6.16}$$

However, since (x_2, y_2) is a maximum point for $v(x,y)$, one must have

$$\frac{\partial^2 v(x_2, y_2)}{\partial x^2} \leq 0. \tag{6.17}$$

From (6.16) and (6.17), the theorem follows by contradiction.

Theorem 6.2. For $b > 0$, let B be the rectangle with vertices $(0,0)$, $(\pi,0)$, (π,b), $(0,b)$, and let R be its interior. If $u = f(x,y)$ is continuous on $R \cup B$ and is a solution of (6.1) on $R \cup L_4^*$, then $f(x,y)$ takes on its minimum value on $L_1 \cup L_2 \cup L_3$.

Proof. The proof follows immediately by applying Theorem 6.1 to the function $-f(x,y)$.

Theorem 6.3. If a boundary-value problem of type H has a solution, then that solution is unique.

Proof. Let $f_1(x,y)$, $f_2(x,y)$ be two solutions of a given boundary-value problem of type H. Then $f(x,y) = f_1(x,y) - f_2(x,y)$ is a solution of (6.1) which is continuous on $R \cup B$. Hence, $f(x,y)$ takes on its maximum and minimum values on $L_1 \cup L_2 \cup L_3$. But, on

$L_1 \cup L_2 \cup L_3$, $f(x,y) = 0$, so that $f(x,y) \equiv 0$ on $R \cup B$. Therefore, $f_1(x,y) \equiv f_2(x,y)$ on $R \cup B$ and the theorem is proved.

Theorem 6.4. If a boundary-value problem of type H has a solution $u = f(x,y)$, then f depends continuously on $g_1(y), g_2(x), g_3(y)$.

Proof. The proof is completely analogous to that of Theorem 5.6 and is a direct consequence of Theorems 6.1 and 6.2.

6.2 Series Solution of Boundary-value Problems of Type H

With the theoretical basis provided by Theorems 6.1 to 6.4, one can now reasonably explore the actual solution of a large class of boundary-value problems of type H. Thus, as described previously, for $b > 0$ let B be the rectangle with vertices $(0,0)$, $(\pi,0)$, (π,b), $(0,b)$, and let R be its interior. Denote the straight-line segments connecting $(0,b)$ and $(0,0)$, $(0,0)$ and $(\pi,0)$, $(\pi,0)$ and (π,b), (π,b) and $(0,b)$ by L_1, L_2, L_3, L_4, respectively. Set L_4 minus its end points $(0,b)$, (π,b) equal to L_4^*. Let $g(x)$, $0 \le x \le \pi$ be continuous and satisfy $g(0) = g(\pi) = 0$. The problem to be explored is that of producing a function $u = f(x,y)$ which

(a) is continuous on $R \cup B$,

(b) satisfies the heat equation

$$u_{xx} = u_y \tag{6.18}$$

on $R \cup L_4^*$, and

(c) satisfies

$$f(0,y) \equiv 0 \quad \text{on } L_1, \tag{6.19}$$

$$f(x,0) \equiv g(x) \text{ on } L_2, \tag{6.20}$$

$$f(\pi,y) \equiv 0, \quad \text{on } L_3. \tag{6.21}$$

Consider first, in a heuristic fashion, a solution of the above problem in the form

$$f(x,y) = X(x)Y(y). \tag{6.22}$$

Substitution of (6.22) into (6.18) implies

$$\frac{d^2X}{dx^2} - \lambda X = 0, \tag{6.23}$$

$$\frac{dY}{dy} - \lambda Y = 0, \tag{6.24}$$

where λ is a constant.

Now if $g(x) \equiv 0$, $0 \le x \le \pi$, then $f(x,y) = 0$ is the unique solution of the problem. Suppose then that $g(x) \not\equiv 0$, $0 \le x \le \pi$, so that neither $X(x)$, $0 \leq x \leq \pi$; $Y(y)$, $0 \leq y \leq b$ is identically zero on its domain of definition. In order to satisfy (6.19) and (6.21), allow

$$X(0) = X(\pi) = 0. \tag{6.25}$$

Subjecting the general solution

$$X(x) = c_1 \sin \sqrt{-\lambda}x + c_2 \cos \sqrt{-\lambda}x \tag{6.26}$$

of (6.23) to conditions (6.25) yields

$$\left. \begin{array}{r} c_1 \sin \sqrt{-\lambda}\pi = 0 \\ c_2 = 0. \end{array} \right\} \tag{6.27}$$

Since $X(x)$ cannot be identically zero, (6.26) and (6.27) imply $\lambda = -n^2$; $n = 1, 2, \ldots$. Depending then on n, the solution $X_n(x)$ of (6.23) subject to (6.25) is

$$X_n(x) = A_n \sin nx, \tag{6.28}$$

where A_n is an arbitrary nonzero constant for $n = 1, 2, \ldots$.

Moreover, $\lambda = -n^2$, $n = 1, 2, \ldots$, implies with respect to (6.24) that

$$Y_n(y) = B_n e^{-n^2 y}, \qquad n = 1, 2, \ldots, \tag{6.29}$$

where each B_n is an arbitrary constant. Hence (6.22), (6.28), and (6.29) suggest consideration of

$$f_n(x,y) = X_n(x)Y_n(y) = A_n B_n (\sin nx) e^{-n^2 y} \tag{6.30}$$

as a solution of (6.18) which satisfies (6.19) and (6.21). Since A_n, B_n are arbitrary constants, set $b_n^* = A_n B_n$ and rewrite (6.30) as

$$f_n(x,y) = b_n^* (\sin nx) e^{-n^2 y}. \tag{6.31}$$

In order to satisfy (6.20), consider

$$f(x,y) = \sum_{n=1}^{\infty} b_n^* (\sin nx) e^{-n^2 y}. \tag{6.32}$$

Then (6.20) and (6.30) imply that on $0 \leq x \leq \pi$

$$g(x) \equiv \sum_{n=1}^{\infty} b_n^* \sin nx,$$

so that the b_n^* must be the coefficients of the Fourier sine series for $g(x)$ and the construction of $f(x,y)$ is complete.

The details involved in then verifying whether or not (6.32) converges, is termwise differentiable, and is actually a solution of the given problem will now be illustrated by two examples.

Example 1. For the boundary-value problem of type H with conditions (6.18) to (6.21), find the unique solution $u = f(x,y)$ on $R \cup B$ which corresponds to $g(x) = \sin x$.

Solution. Since the coefficients of the Fourier sine series for $g(x)$ are, in this case, $b_1^* = 1$, $b_k^* = 0$, $k = 2, 3, \ldots$, then, by (6.32),

$$f(x,y) = (\sin x)e^{-y}. \tag{6.33}$$

That (6.33) is a solution of the problem follows by direct computation. That it is the unique solution of the problem follows from Theorem 6.3.

Example 2. For the boundary-value problem of type H with conditions (6.18) to (6.21), find the unique solution $u = f(x,y)$ on $R \cup B$ which corresponds to $g(x) = 1 - \cos 2x$.

Solution. Since the Fourier sine series for $g(x)$ is

$$1 - \cos 2x \equiv \sum_{n=1}^{\infty} b_n^* \sin nx, \qquad 0 \leq x \leq \pi, \tag{6.34}$$

where

$$\left. \begin{array}{l} b_1^* = \dfrac{16}{3\pi}, \qquad b_2^* = 0 \\[2ex] b_k^* = \dfrac{2}{\pi}\left[\dfrac{1}{k} - \dfrac{\cos k\pi}{k} + \dfrac{\cos(k-2)\pi}{2(k-2)} + \dfrac{\cos(k+2)\pi}{2(k+2)}\right. \\[2ex] \left. \qquad - \dfrac{1}{2(k-2)} - \dfrac{1}{2(k+2)}\right], \qquad k = 3, 4, \ldots, \end{array} \right\} \tag{6.35}$$

then

$$f(x,y) = \sum_{n=1}^{\infty} b_n^*(\sin nx)e^{-n^2 y} \tag{6.36}$$

presents itself as a solution.

As shown following (5.94), (6.34) converges absolutely and uniformly on $0 \leq x \leq \pi$. Moreover, $\{e^{-n^2 y}\}$ is monotonically decreasing with respect to n and uniformly bounded on $R \cup B$. Thus, by Abel's test, the series (6.32) converges uniformly in the two variables x and y together on $R \cup B$. Since $R \cup B$ is closed and bounded, it follows that the series converges to a continuous function, so that $f(x,y)$ is continuous on $R \cup B$.

Moreover, (6.36) implies $f(0,y) \equiv 0$ on L_1, $f(\pi,y) \equiv 0$ on L_3, and by construction $f(x,0) \equiv g(x)$ on L_2.

By Corollary 1 of Theorem 2.6, it can be shown that there exists a positive constant M such that $|b_n^*| < M$; $n = 1, 2, \ldots$. Then, since

$$|-n^2 b_n^* e^{-n^2 y}\sin nx| \leq n^2 M e^{-n^2 y}, \qquad y \geq 0,$$

and since the series $\sum_{n=1}^{\infty} n^2 e^{-n^2 y}$ converges for $y > 0$, so that

$$\sum_{n=1}^{\infty} [-n^2 b_n^* e^{-n^2 y} \sin nx]$$

converges absolutely and uniformly in y, for $y > 0$, it follows that (6.36) can be differentiated with respect to y, for $y > 0$, so that

$$\frac{\partial f}{\partial y} = \sum_{n=1}^{\infty} [-n^2 b_n^* e^{-n^2 y} \sin nx]. \tag{6.37}$$

In a similar fashion, for $0 < x < \pi$,

$$\frac{\partial^2 f}{\partial x^2} = -\sum_{n=1}^{\infty} [n^2 b_n^* e^{-n^2 y} \sin nx]. \tag{6.38}$$

Hence, (6.37) and (6.38) imply that (6.36) satisfies the heat equation on $R \cup L_4^*$.

Finally, Theorem 6.3 implies the uniqueness of (6.36).

It is worth noting that the process of finding $\dfrac{\partial^2 f}{\partial x^2}$, $\dfrac{\partial f}{\partial y}$ in Example 2, above, may be extended easily to the calculation of all partial derivatives of all orders. Hence, by Taylor's theorem, $f(x,y)$ is readily shown to be analytic at all points (x,y) whose coordinates satisfy $0 < x < \pi$, $y > 0$.

EXERCISES

6.1. For the boundary-value problem of type H with conditions (6.18) to (6.21), find, if possible, on $R \cup B$ a solution $u = f(x,y)$ which corresponds to each of the following functions $g(x)$, check each answer, and discuss in each case uniqueness:

(a) $g(x) = \sin 2x$

(b) $g(x) = \sin x + 2 \sin 3x$

(c) $g(x) = \sum_{k=1}^{n} \sin kx$, n a fixed positive integer

(d) $g(x) = \sin^2 x$

(e) $g(x) = 1 - \cos 4x$

(f) $g(x) = -\pi^2 x^2 + 2\pi x^3 - x^4$.

(g) $g(x) = \begin{cases} x, & 0 \le x \le \frac{1}{2}\pi \\ \pi - x, & \frac{1}{2}\pi \le x \le \pi. \end{cases}$

6.2. For the boundary-value problem of type H with conditions (6.2) to (6.4), find, if possible, on $R \cup B$ a solution $u = f(x,y)$ which corresponds to each of the following sets of functions $g_1(y)$, $g_2(x)$, $g_3(y)$, check each answer, and discuss in each case uniqueness:

(a) $g_1(y) = 0$, $g_2(x) = \sin x$, $g_3(y) = \sin y$
(b) $g_1(y) = \sin y$, $g_2(x) = \sin x$, $g_3(y) = \sin y$
(c) $g_1(y) = \sin 3y$, $g_2(x) = 1 - \cos 2x$, $g_3(y) = \sin^2 y$.

6.3. On the plane point set $R \cup B$ described in Sec. 6.1 and displayed in Diagram 6.1, find a continuous function $u = f(x,y)$ which is a solution of the heat equation on $R \cup L_4^*$ and which satisfies

$$f(x,0) = \sin x, \qquad 0 \leq x \leq \pi,$$

$$\frac{\partial f(0,y)}{\partial x} = 0, \qquad 0 < y < b,$$

$$\frac{\partial f(\pi,y)}{\partial x} = 0, \qquad 0 < y < b.$$

6.4. On the plane point set $R \cup B$ described in Sec. 6.1 and displayed in Diagram 6.1, find a continuous function $u = f(x,y)$ which is a solution of

$$4\frac{\partial^2 u}{\partial x^2} = \frac{\partial u}{\partial y}$$

on $R \cup L_4^*$ and which satisfies

$$f(0,y) = 0, \qquad 0 \leq y \leq b,$$
$$f(x,0) = \sin x, \qquad 0 \leq x \leq \pi,$$
$$f(\pi,y) = 0, \qquad 0 \leq y \leq b.$$

6.5. Let B be the rectangle with vertices $(0,0)$, $(1,0)$, $(1,2)$, $(0,2)$, and let R be its interior.   On $R \cup B$ find a continuous function $u = f(x,y)$ which:
(a) satisfies the heat equation for all (x,y) whose coordinates satisfy $0 < x < 1$, $0 < y \leq 2$, and
(b) satisfies

$$f(0,y) = 0, \qquad 0 \leq y \leq 2,$$
$$f(x,0) = \sin 2\pi x, \qquad 0 \leq x \leq 1,$$
$$f(\pi,y) = 0, \qquad 0 \leq y \leq 2.$$

Chapter 7

APPROXIMATE SOLUTION OF
PARTIAL DIFFERENTIAL EQUATIONS

Careful evaluation of the mathematical structure thus far developed reveals numerous difficulties involved in exhibiting the solution of a given partial differential equation problem. Evaluation of a Poisson integral or of Fourier coefficients may involve integrals which are not readily, if at all, expressible in closed form, while existence theorems have indicated that unique solutions exist for a far wider class of problems than those for which explicit solutions have been obtained. In this section, then, methods for *approximating* solutions of partial differential equations will be discussed. A complete mathematical foundation will be provided for the method related to the Dirichlet problem, while appropriately referenced techniques will be offered, usually without proof, for problems connected with the wave and with the heat equations.

7.1 Lattice Points

Definition 7.1. In E^2, let R be a region whose boundary is B. Let $(\bar{x},\bar{y})$ be a point of $R \cup B$. For fixed $h > 0$, let L_h denote the set of all points $(\bar{x} + mh, \bar{y} + nh)$, $m = 0, \pm 1, \pm 2, \ldots, n = 0, \pm 1, \pm 2, \ldots$, contained in $R \cup B$. Then L_h is said to be a *grid*, or a *set of lattice points*, defined on $R \cup B$, and h is said to be the *grid size* or *mesh width*.

Example. Let B be the ellipse whose equation is $9x^2 + 16y^2 = 144$ and let R be its interior. For $h = 1$, $(\bar{x},\bar{y}) = (0,0)$, the corresponding set L_h is the collection of points (0,3), (−2,2), (−1,2), (0,2), (1,2), (2,2), (−3,1), (−2,1), (−1,1), (0,1), (1,1), (2,1), (3,1), (−4,0), (−3,0), (−2,0), (−1,0),

153

$(0,0), (1,0), (2,0), (3,0), (4,0), (-3,-1), (-2,-1), (-1,-1), (0,-1), (1,-1),$
$(2,-1), (3,-1), (-2,-2), (-1,-2), (0,-2), (1,-2), (2,-2), (0,-3),$
which have been labeled 1–35, respectively, in Diagram 7.1.

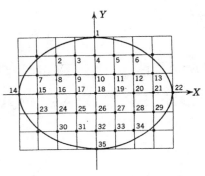

DIAGRAM 7.1

For the purposes of approximation, it will be desirable to divide a given set of lattice points L_h into two disjoint subsets B_h and R_h, where B_h will consist of those points of L_h which, in some convenient sense, lie near B, and will then be called the lattice boundary, while the remaining points of L_h will constitute R_h and will be called the interior lattice points. These concepts are now made precise by means of the following definitions.

Definition 7.2. In E^2, let R be a region whose boundary is B. For fixed $h > 0$, let L_h be a set of lattice points defined on $R \cup B$. Then two points (x_1, y_1), (x_2, y_2) of L_h are said to be *adjacent* if and only if the straight-line segment joining them is contained in $R \cup B$ and

(a) $x_1 = x_2, \quad |y_2 - y_1| = h,$ or

(b) $y_1 = y_2, \quad |x_2 - x_1| = h.$

Example. Let B be the ellipse whose equation is $9x^2 + 16y^2 = 144$, and let R be its interior. For $h = 1$, $(\bar{x}, \bar{y}) = (0,0)$, let the points of L_h be numbered as in Diagram 7.1. Then the pairs of points numbered 1,4; 5,6; 26,32 are pairs of adjacent points, while the pairs numbered 1,5; 1,10; 8,10; 6,24 are not.

Definition 7.3. In E^2, let R be a region whose boundary is B. For fixed $h > 0$, let L_h be a set of lattice points defined on $R \cup B$. Then the *interior* of L_h, denoted by R_h, is the set of all points of L_h which have four adjacent points in L_h.

Definition 7.4. In E^2, let R be a region whose boundary is B. For fixed $h > 0$, let L_h be a set of lattice points defined on $R \cup B$. Then the *boundary* of L_h, denoted by B_h, and called the *lattice boundary*, is defined by $B_h \cup R_h = L_h$, $B_h \cap R_h = \theta$.

Example. Let B be the ellipse whose equation is $9x^2 + 16y^2 = 144$ and let R be its interior. For $h = 1$ and $(\bar{x}, \bar{y}) = (0,0)$, let the resulting set of lattice points L_h be numbered 1–35, as shown in Diagram 7.1. Then R_h consists of those points numbered 4, 8–12, 15–21, 24–28, 32, while B_h consists of those points numbered 1–3, 5–7, 13, 14, 22, 23, 29–31, 33–35.

If $u = f(x,y)$ is the unknown solution of a partial differential equation problem on a plane point set $R \cup B$, then the following notational device will often be convenient. Suppose L_h is a set of lattice points defined on $R \cup B$ and that L_h consists of exactly n points. Number these points in a one-to-one fashion with the integers $1, 2, \ldots, n$. Then, denote the co-ordinates of the point numbered k by (x_k, y_k) and the unknown function $u = f(x,y)$ at (x_k, y_k) by u_k, $k = 1, 2, \ldots, n$.

7.2   Approximate Solution of the Dirichlet Problem

For the Dirichlet problem (Sec. 5.1), a procedure for approximating the unique solution will now be described in four general steps. After an illustrative problem, uniqueness, convergence, and other mathematical questions will be explored.

Liebmann-Gerschgorin-Collatz Approximation Procedure

Step 1. For $(\bar{x}, \bar{y}) \in (R \cup B)$ and for fixed $h > 0$, construct a set of lattice points on $R \cup B$. If L_h consists of n distinct points, number these in a one-to-one fashion with the integers $1, 2, \ldots, n$.

Step 2. At each point of the lattice boundary, approximate the analytical solution $u = f(x,y)$ of the Dirichlet problem as follows. Let (x_1, y_1) be a point of B_h and suppose that it has been assigned the number 1. If (x_1, y_1) is, in addition, a point of B, then set

$$u_1 = g(x_1, y_1). \tag{7.1}$$

If (x_1, y_1) is not a point of B, then at least one of the four points $(x_1 + h, y_1)$, $(x_1, y_1 + h)$, $(x_1 - h, y_1)$, $(x_1, y_1 - h)$ is either not a point of $R \cup B$ or is a point of $R \cup B$ which is not adjacent to (x_1, y_1). Without loss of generality, assume that one such point is $(x_1 + h, y_1)$. Then there exists a point $(x_1 + d, y_1)$, $0 < d < h$, which, say, is numbered 2', which lies on B, and for which the straight-line segment joining (x_1, y_1) and $(x_1 + d, y_1)$ is contained in $R \cup B$. Two cases must then be considered.

CASE 1. Suppose $(x_1 - h, y_1)$ is a point of L_h which is adjacent to (x_1, y_1). Then, as in Diagram 7.2, let $(x_1 - h, y_1)$ be numbered 3

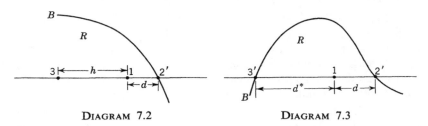

DIAGRAM 7.2   DIAGRAM 7.3

and approximate u_1 by means of linear interpolation, that is, approximate u_1 by

$$u_1 = \frac{hu_{2'} + du_3}{h + d}.$$

However, $u_{2'} = g(x_1 + d, y_1)$, so that the latter relationship may be written

$$u_1 = \frac{du_3 + hg(x_1 + d, y_1)}{d + h}. \tag{7.2}$$

CASE 2. Suppose $(x_1 - h, y_1)$ is either not a point of L_h or is a point of L_h which is not adjacent to (x_1, y_1). Then, as in Diagram 7.3, there exists a point $(x_1 - d^*, y_1)$, $0 < d^* < h$, which, say, is numbered $3'$, which lies on B, and for which the straight-line segment joining (x_1, y_1) to $(x_1 - d^*, y_1)$ is contained entirely in $R \cup B$. By means of linear interpolation, approximate u_1 by

$$u_1 = \frac{du_{3'} + d^*u_{2'}}{d + d^*},$$

or, equivalently, by

$$u_1 = \frac{dg(x_1 - d^*, y_1) + d^*g(x_1 + d, y_1)}{d + d^*}. \tag{7.3}$$

Note that the application of any interpolation scheme to approximate $u = f(x, y)$ on B_h is reasonable since $f(x, y)$ is continuous on $R \cup B$.

Step 3. At each point of R_h proceed as follows. Let (x_1, y_1) be a point of R_h, and let it be numbered 1, as in Diagram 7.4. Let the four points $(x_1 + h, y_1)$, $(x_1, y_1 + h)$, $(x_1 - h, y_1)$, $(x_1, y_1 - h)$ of L_h be numbered 2, 3, 4, 5, respectively. Then set

$$-4u_1 + u_2 + u_3 + u_4 + u_5 = 0. \tag{7.4}$$

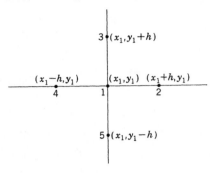

DIAGRAM 7.4

Note that (7.4) is a reasonable replacement of the potential equation, for a necessary and sufficient condition that a function be harmonic on R is that it possess the mean value property on R, while (7.4) implies:

$$u_1 = \frac{u_2 + u_3 + u_4 + u_5}{4},$$

which, in turn implies that u_1 is the mean value of u_2, u_3, u_4, u_5. A precise development of (7.4) will be given later.

Step 4. Application of Steps 1, 2, and 3, above, and use of the subscript notation yield a system of n linear algebraic equations in the n unknowns u_1, u_2, $\ldots$, u_n. The final step, then, is to solve this linear algebraic system to yield an approximation of the analytical solution $u = f(x,y)$ at each point of L_h.

Example. Let B be the triangle with vertices $(-4,0)$, $(4,0)$, $(0,3)$, and let R be its interior. Let $g(x,y) \equiv 16 - x^2$ on the side of B which connects $(-4,0)$, and $(4,0)$, and let it be zero on the remainder of B. Set $(\bar{x},\bar{y}) = (0,0)$ and $h = 1$. As shown in Diagram 7.5, the points of L_h are then

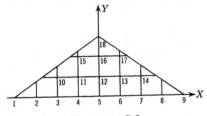

DIAGRAM 7.5

$(-4,0)$, $(-3,0)$, $(-2,0)$, $(-1,0)$, $(0,0)$, $(1,0)$, $(2,0)$, $(3,0)$, $(4,0)$, $(-2,1)$, $(-1,1)$, $(0,1)$, $(1,1)$, $(2,1)$, $(-1,2)$, $(0,2)$, $(1,2)$, $(0,3)$ and are numbered, respectively, 1–18. Considering each point in order and applying the appropriate step of the Liebmann–Gerschgorin–Collatz procedure yields the eighteen equations:

$$u_1 = 0; \qquad u_{10} = \frac{\frac{2}{3}u_{11} + 0}{1 + \frac{2}{3}}$$

$$u_2 = 7; \qquad -4u_{11} + u_{12} + u_{15} + u_{10} + u_4 = 0$$

$$u_3 = 12; \qquad -4u_{12} + u_{13} + u_{16} + u_{11} + u_5 = 0$$

$$u_4 = 15; \qquad -4u_{13} + u_{14} + u_{17} + u_{12} + u_6 = 0$$

$$u_5 = 16; \qquad u_{14} = \frac{\frac{2}{3}u_{13} + 0}{1 + \frac{2}{3}}$$

$$u_6 = 15; \qquad u_{15} = \frac{\frac{1}{3}u_{16} + 0}{1 + \frac{1}{3}}$$

$$u_7 = 12; \qquad -4u_{16} + u_{17} + u_{18} + u_{15} + u_{12} = 0$$

$$u_8 = 7; \qquad u_{17} = \frac{\frac{1}{3}u_{16} + 0}{1 + \frac{1}{3}}$$

$$u_9 = 0; \qquad u_{18} = 0,$$

the unique solution of which is $u_1 = 0$, $u_2 = 7$, $u_3 = 12$, $u_4 = 15$, $u_5 = 16$, $u_6 = 15$, $u_7 = 12$, $u_8 = 7$, $u_9 = 0$, $u_{10} = 340/131$, $u_{11} = 850/131$, $u_{12} = 1022/131$, $u_{13} = 850/131$, $u_{14} = 340/131$, $u_{15} = 73/131$, $u_{16} = 292/131$, $u_{17} = 73/131$, $u_{18} = 0$.

Note of course that other approximate solutions could have been generated by performing vertical rather than horizontal interpolation at any combination of the points numbered 10, 14, 15, and 17.

7.3 Mathematical Basis of the Liebmann-Gerschgorin-Collatz Procedure

It will now be shown that the Liebmann-Gerschgorin-Collatz method is a reasonable technique of approximation in that
   (a) if the points (x_1, y_1), $(x_1 + h, y_1)$, $(x_1, y_1 + h)$, $(x_1 - h, y_1)$, $(x_1, y_1 - h)$, numbered 1, 2, 3, 4, 5, respectively, as in Diagram 7.4 are elements of R, then (7.4) can be derived rigorously,
   (b) the solution of the resulting linear algebraic system always exists and is unique, and
   (c) the approximate solution converges to the analytical solution as the mesh width h converges to zero.

Consider then, first, the five points (x_1, y_1), $(x_1 + h, y_1)$, $(x_1, y_1 + h)$, $(x_1 - h, y_1)$, $(x_1, y_1 - h)$, numbered 1, 2, 3, 4, and 5, respectively, as in Diagram 7.4.   Define the difference δ by

$$\delta = (\alpha_1 u_1 + \alpha_2 u_2 + \alpha_3 u_3 + \alpha_4 u_4 + \alpha_5 u_5) - \left[\frac{\partial^2 f(x_1, y_1)}{\partial x^2} + \frac{\partial^2 f(x_1, y_1)}{\partial y^2} \right], \quad (7.5)$$

where α_1, α_2, α_3, α_4, α_5 are constants to be determined and $u = f(x, y)$ is the analytical solution of the Dirichlet problem.   Since (x_1, y_1), $(x_1 + h, y_1)$, $(x_1, y_1 + h)$, $(x_1 - h, y_1)$, $(x_1, y_1 - h)$ are assumed to be elements of R and since $u = f(x, y)$ is analytic on R, it readily follows that

$$u_2 = u(x_1, y_1) + h \frac{\partial f(x_1, y_1)}{\partial x} + \frac{h^2}{2} \frac{\partial^2 f(x_1, y_1)}{\partial x^2} + O(h^3) \qquad (7.6)$$

$$u_3 = u(x_1, y_1) + h \frac{\partial f(x_1, y_1)}{\partial y} + \frac{h^2}{2} \frac{\partial^2 f(x_1, y_1)}{\partial y^2} + O(h^3) \qquad (7.7)$$

$$u_4 = u(x_1, y_1) - h \frac{\partial f(x_1, y_1)}{\partial x} + \frac{h^2}{2} \frac{\partial^2 f(x_1, y_1)}{\partial x^2} + O(h^3) \qquad (7.8)$$

$$u_5 = u(x_1, y_1) - h \frac{\partial f(x_1, y_1)}{\partial y} + \frac{h^2}{2} \frac{\partial^2 f(x_1, y_1)}{\partial y^2} + O(h^3). \qquad (7.9)$$

Substitution of (7.6) to (7.9) into (7.5) yields

$$\delta = u_1(\alpha_1 + \alpha_2 + \alpha_3 + \alpha_4 + \alpha_5) + \frac{\partial f(x_1, y_1)}{\partial x} (h\alpha_2 - h\alpha_4)$$

$$+ \frac{\partial f(x_1, y_1)}{\partial y} (h\alpha_3 - h\alpha_5) + \frac{\partial^2 f(x_1, y_1)}{\partial x^2} \left(\frac{h^2}{2} \alpha_2 + \frac{h^2}{2} \alpha_4 - 1 \right)$$

$$+ \frac{\partial^2 f(x_1, y_1)}{\partial y^2} \left(\frac{h^2}{2} \alpha_3 + \frac{h^2}{2} \alpha_5 - 1 \right)$$

$$+ \alpha_2 O(h^3) + \alpha_3 O(h^3) + \alpha_4 O(h^3) + \alpha_5 O(h^3). \qquad (7.10)$$

Setting

$$\alpha_1 + \alpha_2 + \alpha_3 + \alpha_4 + \alpha_5 = 0$$

$$h\alpha_2 - h\alpha_4 = 0$$

$$h\alpha_3 - h\alpha_5 = 0$$

$$\frac{h^2}{2} \alpha_2 + \frac{h^2}{2} \alpha_4 - 1 = 0$$

$$\frac{h^2}{2} \alpha_3 + \frac{h^2}{2} \alpha_5 - 1 = 0$$

implies $\alpha_1 = -4/h^2$, $\alpha_2 = \alpha_3 = \alpha_4 = \alpha_5 = 1/h^2$, so that (7.5) and (7.10) imply

$$\frac{-4u_1 + u_2 + u_3 + u_4 + u_5}{h^2} - \left[\frac{\partial^2 f(x_1,y_1)}{\partial x^2} + \frac{\partial^2 f(x_1,y_1)}{\partial y^2}\right] \equiv 0(h),$$

or, equivalently,

$$\frac{-4u_1 + u_2 + u_3 + u_4 + u_5}{h^2} + 0(h) \equiv \frac{\partial^2 f(x_1,y_1)}{\partial x^2} + \frac{\partial^2 f(x_1,y_1)}{\partial y^2}. \quad (7.11)$$

Hence the harmonic property of $f(x,y)$ at (x_1,y_1) implies

$$\frac{-4u_1 + u_2 + u_3 + u_4 + u_5}{h^2} + 0(h) \equiv 0. \quad (7.12)$$

Finally, discarding the terms $0(h)$ in (7.12) leads to the approximation

$$\frac{-4u_1 + u_2 + u_3 + u_4 + u_5}{h^2} = 0$$

or, equivalently, to

$$-4u_1 + u_2 + u_3 + u_4 + u_5 = 0,$$

and the derivation of (7.4) is complete.

Note that (7.4) is called a *finite-difference analogue* of the potential equation.

Further examination of the theoretical basis of the Liebmann-Gerschgorin-Collatz procedure will necessitate a careful delineation between the analytical solution of the Dirichlet problem and the approximate solution under consideration. Hence, in the remainder of the discussion, let $u = f(x,y)$ represent the analytical solution of the problem and let $U = U(x,y)$ represent the approximate solution. Of course, $U = U(x,y)$ is defined only on L_h and, by (7.1) to (7.4), one must have

(a) if (x_1,y_1) is a point of $B_h \cap B$, then

$$U(x_1,y_1) = g(x_1,y_1) \quad (7.13)$$

(b) if (x_1,y_1) is numbered 1, is a point of B_h but is not a point of B, and if the arrangement of the points numbered $2',3$ follows the description of Step 2, Case 1, Sec. 7.2, then

$$U(x_1,y_1) = \frac{dU(x_1 - h, y_1) + hg(x_1 + d, y_1)}{d + h} \quad (7.14)$$

(c) if (x_1, y_1) is numbered 1, is a point of B_h but is not a point of B, and if the arrangement of the points numbered $2', 3'$ follows the description of Case 2, Step 2, Sec. 7.2, then

$$U(x_1, y_1) = \frac{dg(x_1 - d^*, y_1) + d^*g(x_1 + d, y_1)}{d + d^*} \qquad (7.15)$$

(d) if (x_1, y_1) is a point of R_h, then

$$-4U(x_1, y_1) + U(x_1 + h, y_1) + U(x_1, y_1 + h)$$
$$+ U(x_1 - h, y_1) + U(x_1, y_1 - h) = 0. \qquad (7.16)$$

Theorem 7.1. The solution of the system of linear algebraic equations generated by the Liebmann-Gerschgorin-Collatz technique always exists and is unique.

Proof. Consider the linear system which results if $g(x, y) \equiv 0$ on B. This system is homogeneous and always has the zero vector for a solution. The theorem will therefore be valid if it can be shown that in this case the zero vector is the only solution of the system, and this will now be done.

Suppose there exists a nontrivial solution $U(x, y)$ on L_h for the homogeneous system. Then, at some point $(x_k, y_k) \in L_h$, $U(x_k, y_k) \neq 0$. Without loss of generality, assume $U(x_k, y_k) < 0$. Let the minimum value of $U(x, y)$ on L_h be $-M$ and let $(\bar{x}_1, \bar{y}_1)$ be any point of L_h such that

$$U(\bar{x}_1, \bar{y}_1) = -M < 0. \qquad (7.17)$$

Then, for all $(x, y) \in L_h$,

$$U(\bar{x}_1, \bar{y}_1) = -M \leq U(x, y). \qquad (7.18)$$

If $(\bar{x}_1, \bar{y}_1) \in B_h$, then:
(a) (7.13) implies $U(\bar{x}_1, \bar{y}_1) = 0$, which contradicts (7.17),
(b) (7.14) implies $U(\bar{x}_1, \bar{y}_1) > U(\bar{x}_1 - h, \bar{y}_1)$, which contradicts (7.18),
(c) (7.15) implies $U(\bar{x}_1, \bar{y}_1) = 0$, which contradicts (7.17).
Thus $(\bar{x}_1, \bar{y}_1)$ *cannot* be an element of B_h.

If $(\bar{x}_1, \bar{y}_1) \in R_h$, then only two possibilities exist and each will be considered in turn:
(a') Suppose $U(\bar{x}_1 + h, \bar{y}_1)$, $U(\bar{x}_1, \bar{y}_1 + h)$, $U(\bar{x}_1 - h, \bar{y}_1)$, $U(\bar{x}_1, \bar{y}_1 - h)$ are all equal. Then it follows from (7.16) that $U(\bar{x}_1, \bar{y}_1) = U(\bar{x}_1 + h, \bar{y}_1)$. But this argument may be repeated a finite number of times to yield that

$$-M = U(\bar{x}_1, \bar{y}_1) = U(\bar{x}_1 + mh, \bar{y}_1),$$

where $(\bar{x}_1 + mh, \bar{y}) \in B_h$. However, as shown above, $U(x,y)$ cannot equal $-M$ at a point of B_h, and a contradiction has been reached.

(b') Suppose at least two of $U(\bar{x}_1 + h, \bar{y}_1)$, $U(\bar{x}_1, \bar{y}_1 + h)$, $U(\bar{x}_1 - h, \bar{y}_1)$, $U(\bar{x}_1, \bar{y}_1 - h)$ are not equal. Then there exists, amongst these, a minimum. Without loss of generality, assume that $U(\bar{x}_1 + h, \bar{y}_1)$ is that minimum. Hence

$$U(\bar{x}_1, \bar{y}_1 + h) = U(\bar{x}_1 + h, \bar{y}_1) + \gamma_1, \tag{7.19}$$

$$U(\bar{x}_1 - h, \bar{y}_1) = U(\bar{x}_1 + h, \bar{y}_1) + \gamma_2, \tag{7.20}$$

$$U(\bar{x}_1, \bar{y}_1 - h) = U(\bar{x}_1 + h, \bar{y}_1) + \gamma_3, \tag{7.21}$$

where γ_1, γ_2, γ_3 are nonnegative and at least one is positive. Substitution of (7.19) to (7.21) into (7.15) implies

$$U(\bar{x}_1 + h, \bar{y}_1) < U(\bar{x}_1, \bar{y}_1),$$

thus contradicting (7.18). Hence, (b') is not possible.

Thus, in each possible case, the assumption that a nontrivial solution $U(x,y)$ exists has led to a contradiction and the theorem is proved.

Finally, consider the question of convergence of the approximate solution to the analytical solution.

Definition 7.5. Provided each of the indicated function values exists, let

$$\lambda[v(x,y)] = \frac{1}{h^2}\{-4v(x,y) + v(x + h, y) + v(x, y + h)$$

$$+ v(x - h, y) + v(x, y - h)\}. \tag{7.22}$$

Example 1. On E^2, let $v(x,y) = x - y$. Then $\lambda[x - y] = 0$.

Example 2. On E^2, let $v(x,y) = x^2 + y^2$. Then $\lambda[x^2 + y^2] = 4$.

Example 3. On E^2, let $v(x,y) = x^3 + y$. Then $\lambda[x^3 + y] = 6x$.

Example 4. On E^2, let $v(x,y) = x^3 + y^3$. Then $\lambda[x^3 + y^3] = 6(x + y)$.

Lemma 7.1. If c_1, c_2 are arbitrary constants, then, provided all the indicated function values exist,

$$\lambda[c_1 v_1(x,y) + c_2 v_2(x,y)] = c_1 \lambda[v_1(x,y)] + c_2 \lambda[v_2(x,y)]. \tag{7.23}$$

Proof.

$$\lambda[c_1 v_1(x,y) + c_2 v_2(x,y)] \equiv \frac{1}{h^2}\{-4[c_1 v_1(x,y) + c_2 v_2(x,y)]$$

$$+ [c_1 v_1(x + h, y) + c_2 v_2(x + h, y)]$$

$$+ [c_1 v_1(x, y + h) + c_2 v_2(x, y + h)]$$

$$+ [c_1 v_1(x - h, y) + c_2 v_2(x - h, y)]$$

$$+ [c_1 v_1(x, y - h) + c_2 v_2(x, y - h)]\}$$

$$\equiv c_1[-4v_1(x,y) + v_1(x + h, y)$$
$$+ v_1(x, y + h) + v_1(x - h, y)$$
$$+ v_1(x, y - h)] + c_2[-4v_2(x,y)$$
$$+ v_2(x + h, y) + v_2(x, y + h)$$
$$+ v_2(x - h, y) + v_2(x, y - h)]$$
$$\equiv c_1\lambda[v_1(x,y)] + c_2\lambda[v_2(x,y)].$$

Lemma 7.2. Let R be a region, B its boundary, and L_h a set of lattice points defined on $R \cup B$. Let $v(x,y)$ be defined on L_h. If $\lambda[v(x,y)] \le 0$ on R_h and $v(x,y) \ge 0$ on B_h, then $v(x,y) \ge 0$ on R_h.

Proof. Suppose $v(x,y) < 0$ at some point of R_h. There exists a point $(x_1,y_1) \in R_h$ at which $v(x,y)$ is a minimum. Hence

$$v(x_1,y_1) < 0 \tag{7.24}$$

and for all $(x,y) \in R_h$

$$v(x_1,y_1) \le v(x,y). \tag{7.25}$$

Since $\lambda[v(x,y)] \le 0$ by assumption, it follows from (7.22) that for all $(x,y) \in R_h$

$$v(x,y) \ge \tfrac{1}{4}[v(x + h, y) + v(x, y + h) + v(x - h, y) + v(x, y - h)]. \tag{7.26}$$

But (7.24) to (7.26) imply, by the techniques displayed in Theorem 7.1, that $v(x,y) < 0$ at some point of B_h. Thus, by contradiction, the proof is complete.

Lemma 7.3. Let R be a region, B its boundary, and L_h a set of lattice points defined on $R \cup B$. Let $v_1(x,y)$, $v_2(x,y)$ be defined on L_h. If $-|\lambda[v_1(x,y)]| \ge \lambda[v_2(x,y)]$ on R_h and $|v_1(x,y)| \le v_2(x,y)$ on B_h, then $|v_1(x,y)| \le v_2(x,y)$ on R_h.

Proof. $v_2(x,y) - v_1(x,y) \ge 0$ on B_h and $\lambda[v_2(x,y) - v_1(x,y)] = \lambda[v_2(x,y)] - \lambda[v_1(x,y)]$ on R_h. Since $\lambda[v_2(x,y)] + |\lambda[v_1(x,y)]| \le 0$ on R_h, then $\lambda[v_2(x,y)] - \lambda[v_1(x,y)] \le \lambda[v_2(x,y)] + |\lambda[v_1(x,y)]| \le 0$ on R_h. Hence, $\lambda[v_2(x,y) - v_1(x,y)] \le 0$ on R_h. Thus, by Lemma 7.2, $v_2(x,y) - v_1(x,y) \ge 0$ on R_h.

Also, $v_2(x,y) + v_1(x,y) \ge v_2(x,y) - |v_1(x,y)| \ge 0$ on B_h and $\lambda[v_2(x,y) + v_1(x,y)] = \lambda[v_2(x,y)] + \lambda[v_1(x,y)] \le 0$ on R_h. Hence, $v_2(x,y) + v_1(x,y) \ge 0$ on R_h.

Finally, the conclusions $v_2(x,y) + v_1(x,y) \ge 0$ on R_h and $v_2(x,y) - v_1(x,y) \ge 0$ on R_h imply $v_2(x,y) \ge |v_1(x,y)|$ on R_h, and the lemma is proved.

Lemma 7.4. Let R be a bounded region, B its boundary, and L_h a set of lattice points defined on $R \cup B$. Let $v(x,y)$ be defined on L_h.

If $|\lambda[v(x,y)]| \leq A_1$ on R_h, $|v(x,y)| \leq A_2$ on B_h, and r is the radius of any circle which contains $R \cup B$ in its interior, then on R_h

$$|v(x,y)| \leq \tfrac{1}{4}A_1 r^2 + A_2. \qquad (7.27)$$

Proof.   Let

$$(x - a)^2 + (y - b)^2 = r^2$$

be the equation of a circle with center (a,b) and radius r which contains $R \cup B$ in its interior.   Consider

$$w(x,y) = \left[\tfrac{1}{4}A_1 r^2\left\{1 - \frac{(x - a)^2 + (y - b)^2}{r^2}\right\} + A_2\right].$$

Direct calculation yields $\lambda[w(x,y)] = -A_1$.   Also, $w(x,y) \geq A_2$ on B_h.   Since $|\lambda[v(x,y)]| \leq A_1$ on R_h, by assumption, and $\lambda[w(x,y)] = -A_1$, it follows that on R_h $|\lambda[v(x,y)]| \leq -\lambda[w(x,y)]$.   Also, $|v(x,y)| \leq A_2$ on B_h and $w(x,y) \geq A_2$ on B_h, so that $w(x,y) \geq |v(x,y)|$ on B_h.   Hence, $-|\lambda[v(x,y)]| \geq \lambda[w(x,y)]$ on R_h and $w(x,y) \geq |v(x,y)|$ on B_h.   By Lemma 7.3, then, $w(x,y) \geq |v(x,y)|$ on R_h, or, on R_h

$$|v(x,y)| \leq w(x,y) \leq \tfrac{1}{4}A_1 r^2 + A_2,$$

and the lemma is proved.

Lemma 7.5.   Let $u = f(x,y)$ be the solution of the Dirichlet problem and let $U(x,y)$ be an approximate solution found by the Liebmann-Gershgorin-Collatz method.   If $f(x,y) \in C^4$ on $R \cup B$ and if (x_i,y_i) is an arbitrary point of R_h, then

$$|U(x_i,y_i) - f(x_i,y_i)| \leq \mu + \frac{r^2 h^2 M_4}{24}, \qquad (7.28)$$

where

$$\mu = \max_{B_h} |U(x,y) - f(x,y)|, \qquad (7.29)$$

$$M_4 = \operatorname*{lub}_{R \cup B}\left[\max\left\{\left|\frac{\partial^4 f}{\partial x^4}\right|, \left|\frac{\partial^4 f}{\partial y^4}\right|\right\}\right], \qquad (7.30)$$

and r is the radius of any circle which contains $R \cup B$ in its interior.

Proof.   Note first that μ, defined by (7.29), exists since $f(x,y)$ is continuous on $R \cup B$ and since $U(x,y)$ is defined on only a finite number of points.   Also M_4, defined by (7.30), exists because, by assumption, $f(x,y) \in C^4$ on $R \cup B$.

Let $(x_i,y_i) \in R_h$ and consider

$$Q = \lambda[f(x_i,y_i)]. \qquad (7.31)$$

Substitution of finite Taylor expansions for $f(x_i+h, y_i)$, $f(x_i, y_i+h)$, $f(x_i - h, y_i)$, $f(x_i, y_i - h)$ into $\lambda[f(x_i,y_i)]$ and use of the fact that $u = f(x,y)$ is harmonic at (x_i,y_i) yields

$$Q = \frac{h^2}{4!}\left[\frac{\partial^4 f(\xi_1,y_i)}{\partial x^4} + \frac{\partial^4 f(\xi_2,y_i)}{\partial x^4} + \frac{\partial^4 f(x_i,\eta_1)}{\partial y^4} + \frac{\partial^4 f(x_i,\eta_2)}{\partial y^4}\right], \quad (7.32)$$

where $x_i < \xi_1 < x_i + h, x_i - h < \xi_2 < x_i, y_i < \eta_1 < y_i + h, y_i - h < \eta_2 < y_i$. Hence,

$$|Q| \leq \frac{h^2 M_4}{6}, \quad (7.33)$$

where M_4 is defined by (7.30). Moreover, from Lemma 7.1 and Eqs. (7.16) and (7.33)

$$|\lambda[U(x_i,y_i) - f(x_i,y_i)]| = |\lambda[U(x_i,y_i)] - \lambda[f(x_i,y_i)]|$$

$$= |\lambda[f(x_i,y_i)]| = |Q| \leq \frac{h^2 M_4}{6}. \quad (7.34)$$

Further, by (7.29), on B_h

$$|U(x,y) - f(x,y)| \leq \mu. \quad (7.35)$$

Thus (7.34) and (7.35) imply, by Lemma 7.4, that on R_h

$$|U(x_i,y_i) - f(x_i,y_i)| \leq \mu + \frac{h^2 r^2 M_4}{24},$$

where r is the radius of any circle which contains $R \cup B$ in its interior, and the proof is complete.

Lemma 7.6. Let $u = f(x,y)$ be the solution of the Dirichlet problem and let $U(x,y)$ be an approximate solution found by the Liebmann-Gerschgorin-Collatz procedure. If $f(x,y) \in C^4$ on $R \cup B$ and if (x_i,y_i) is an arbitrary point of R_h, then

$$|U(x_i,y_i) - f(x_i,y_i)| \leq \frac{r^2 h^2 M_4}{12} + 3h^2 M_2, \quad (7.36)$$

where

$$M_j = \operatorname*{lub}_{R \cup B}\left\{\max\left[\left|\frac{\partial^j f}{\partial x^j}\right|, \left|\frac{\partial^j f}{\partial y^j}\right|\right]\right\}, \quad j = 2, 4 \quad (7.37)$$

and r is the radius of any circle which contains $R \cup B$ in its interior.

Proof. Suppose first that (x_j,y_j) is a point of B_h. Then three cases must be considered:

CASE 1. If $(x_j,y_j) \in (B_h \cap B)$, then

$$|U(x_j,y_j) - f(x_j,y_j)| = |g(x_j,y_j) - g(x_j,y_j)| = 0. \quad (7.38)$$

CASE 2. If (x_j, y_j) is not also a point of B, but is a point of the type discussed in Step 2, Case 1, Sec. 7.2, and is as shown in Diagram 7.2, then

$$f(x_j - h, y_j) = f(x_j + d, y_j) - (h + d)\frac{\partial f(x_j + d, y_j)}{\partial x}$$

$$+ \frac{(h + d)^2}{2!}\frac{\partial^2 f(\xi_1, y_j)}{\partial x^2}, \qquad x_j - h < \xi_1 < x_j + d \quad (7.39)$$

$$f(x_j, y_j) = f(x_j + d, y_j) - d\frac{\partial f(x_j + d, y_j)}{\partial x} + \frac{d^2}{2}\frac{\partial^2 f(\xi_2, y_j)}{\partial x^2},$$

$$x_j < \xi_2 < x_j + d. \quad (7.40)$$

Then, (7.39) implies

$$-d\frac{\partial f(x_j + d, y_j)}{\partial x} = -\frac{d}{h + d}f(x_j + d, y_j) + \frac{d}{h + d}f(x_j - h, y_j)$$

$$- \frac{d(h + d)}{2}\frac{\partial^2 f(\xi_1, y_j)}{\partial x^2}. \quad (7.41)$$

Substitution of (7.41) into (7.40) yields

$$f(x_j, y_j) = \frac{h}{h + d}f(x_j + d, y_j) + \frac{d}{h + d}f(x_j - h, y_j)$$

$$+ \frac{d^2}{2}\frac{\partial^2 f(\xi_2, y_j)}{\partial x^2} - \frac{d(h + d)}{2}\frac{\partial^2 f(\xi_1, y_j)}{\partial x^2}. \quad (7.42)$$

From (7.14), however, since $g(x_j + d, y_j) = f(x_j + d, y_j)$,

$$U(x_j, y_j) = \frac{d}{h + d}U(x_j - h, y_j) + \frac{h}{d + h}f(x_j + d, y_j). \quad (7.43)$$

Hence, (7.42) and (7.43) imply

$$U(x_j, y_j) - f(x_j, y_j) = \frac{d}{h + d}[U(x_j - h, y_j) - f(x_j - h, y_j)]$$

$$- \frac{d^2}{2}\frac{\partial^2 f(\xi_2, y_j)}{\partial x^2} + \frac{d(h + d)}{2}\frac{\partial^2 f(\xi_1, y_j)}{\partial x^2},$$

so that

$$|U(x_j, y_j) - f(x_j, y_j)| \le \tfrac{1}{2}|U(x_j - h, y_j) - f(x_j - h, y_j)| + \tfrac{3}{2}h^2 M_2, \quad (7.44)$$

where M_2 is defined by (7.37).

CASE 3.   If (x_j,y_j) is not also a point of B, but is a point of the type discussed in Step 2, Case 2, Sec. 7.2, and is as shown in Diagram 7.3, then an argument similar to that given above, but with

$$U(x_j,y_j) = \frac{d}{h+d}f(x_j - h, y_j) + \frac{h}{h+d}f(x_j + d, y_j) \qquad (7.45)$$

replacing (7.43), implies

$$|U(x_j,y_j) - f(x_j,y_j)| < \frac{3h^2 M_2}{2}. \qquad (7.46)$$

Hence, if $(x_j,y_j) \in B_h$, then Cases 1, 2, and 3, above, imply by results (7.38), (7.44), and (7.46) that

$$|U(x_j,y_j) - f(x_j,y_j)| \leq \tfrac{1}{2}|U(x_j - h, y_j) - f(x_j - h, y_j)| + \tfrac{3}{2}h^2 M_2. \qquad (7.47)$$

If $(x_j - h, y_j) \in B_h$, then (7.47) implies

$$|U(x_j,y_j) - f(x_j,y_j)| \leq \frac{\mu}{2} + \frac{3h^2 M_2}{2}, \qquad (7.48)$$

while if $(x_j - h, y_j) \in R_h$, then (7.28) and (7.47) imply

$$|U(x_j,y_j) - f(x_j,y_j)| \leq \frac{\mu}{2} + \frac{r^2 h^2 M_4}{48} + \frac{3h^2 M_2}{2}, \qquad (7.49)$$

so that from (7.48) and (7.49), it follows that, *in any case*,

$$|U(x_j,y_j) - f(x_j,y_j)| \leq \frac{\mu}{2} + \frac{r^2 h^2 M_4}{48} + \frac{3h^2 M_2}{2}.$$

However, since (x_j,y_j) is an arbitrary point of B_h, this latter inequality implies

$$\mu \leq \frac{\mu}{2} + \frac{r^2 h^2 M_4}{48} + \frac{3h^2 M_2}{2},$$

or, equivalently, that

$$\mu \leq \frac{r^2 h^2 M_4}{24} + 3h^2 M_2. \qquad (7.50)$$

Substitution of (7.50) into (7.28) implies then that if $(x_i,y_i) \in R_h$,

$$|U(x_i,y_i) - f(x_i,y_i)| \leq \frac{r^2 h^2 M_4}{12} + 3h^2 M_2,$$

and the lemma is proved.

Theorem 7.2.   Let $u = f(x,y)$ be the solution of the Dirichlet problem and let $U(x,y)$ be an approximate solution found by the Liebmann-Gerschgorin-Collatz technique.   If $f(x,y) \in C^4$ on $R \cup B$, then $U(x,y)$ converges to $f(x,y)$ as h converges to zero.

Proof.   The proof at points of R_h follows directly from (7.36), while the proof at points of B_h follows directly from (7.29) and (7.50).

Some final remarks are in order concerning convergence Theorem 7.2. First, note that if one calls $|U(x,y) - f(x,y)|$ the error in the approximate solution $U(x,y)$, then (7.36) is an *error bound*. However, (7.36) is a "weak" error bound in the sense that the terms M_2, M_4 are known to exist, but, in general, cannot be evaluated. Second, note that if it is known that the analytical solution $u = f(x,y)$ is harmonic in some region R', where $(R \cup B) \subset R'$, then, by Exercise 5.10, one would have that $f(x,y) \in C^4$ on $R \cup B$ and this assumption would not be necessary in the statement of Theorem 7.2.

7.4 Approximation Methods for Cauchy Problem I, Cauchy Problem II, the Mixed-type Problem, and the Boundary-value Problem of Type H

The techniques to be described now are of value in the solution of problems associated with the wave and heat equations. For a discussion of the fundamental theories and the various limitations, the reader should consult such sources as Collatz [6], Crank and Nicolson [10], Hartree [20], John [26], Kantorovich and Krylov [27], Milne [34], and Richtmyer [39].

Method I. For Cauchy Problem I (Sec. 4.2), the numerical value of the solution $u = f(x,y)$ at any point $(\bar{x}, \bar{y})$ can be calculated from (4.27). However, if $\int q(t)\, dt$ is not readily expressible in closed form, then $\int_{\bar{x}-\bar{y}}^{\bar{x}+\bar{y}} q(t)\, dt$ may be approximated by any method of numerical integration, as, for example, Simpson's rule, thus leading to an approximation of $f(\bar{x}, \bar{y})$.

Example. Let N be the set of real numbers. Let $u = f(x,y)$ be the solution on E^2 of the homogeneous wave equation (4.2) which satisfies the conditions

$$f(x,0) = 2x, \qquad x \in N$$

$$\frac{\partial f(x,0)}{\partial y} = \begin{cases} \sqrt{4 + x^3}, & x > 0 \\ 2, & x \le 0. \end{cases}$$

Then approximate $f(3,3)$.

Solution. By (4.27),

$$f(x,y) = \frac{2(x + y) + 2(x - y)}{2} + \frac{1}{2}\int_{x-y}^{x+y} \sqrt{4 + t^3}\, dt.$$

Hence,

$$f(3,3) = 6 + \frac{1}{2}\int_0^6 \sqrt{4 + t^3}\, dt. \tag{7.51}$$

By dividing the interval $0 \leq t \leq 6$ into six equal parts and applying Simpson's rule, one finds

$$\int_0^6 \sqrt{4 + t^3}\, dt \sim \tfrac{1}{3}[\sqrt{4 + 0^3} + 4\sqrt{4 + 1^3} + 2\sqrt{4 + 2^3} + 4\sqrt{4 + 3^3}$$
$$+ 2\sqrt{4 + 4^3} + 4\sqrt{4 + 5^3} + \sqrt{4 + 6^3}] \sim 38.90.$$

Hence, from (7.51),

$$f(3,3) \sim 6 + \tfrac{1}{2}(38.90) \sim 25.5.$$

Method II. For Cauchy Problem I, suppose that $u = f(x, y)$ is the analytical solution, $(\bar{x},\bar{y})$ is a fixed point, and that one wishes to approximate $f(\bar{x},\bar{y})$. Without loss of generality, consider $\bar{y} > 0$, since the case $\bar{y} < 0$ follows in a completely analogous fashion. For $\bar{c}$ a fixed, positive integer, set $\bar{y}/\bar{c} = h$. Let B denote the right triangle with vertices $(\bar{x},\bar{y})$, $(\bar{x} - \bar{y}, 0)$, $(\bar{x} + \bar{y}, 0)$ and let R denote the interior of B. On $R \cup B$, let L_h denote the set of lattice points, $(\bar{x} + mh, \bar{y} + nh)$.

For convenience, the points of L_h will be categorized as follows. The points of L_h of the form $(\bar{x} + mh, rh)$, r fixed and equal to one of $0, 1, 2, \ldots, \bar{c}; m = 0, \pm 1, \pm 2, \ldots, \pm\bar{c}$, are called the rth row of L_h.

The method of approximation proceeds by determining $f(x,y)$ at each point of L_h of the form $(\bar{x} + mh, 0)$ by means of the relationship

$$f(\bar{x} + mh,0) = p(\bar{x} + mh), \qquad m = 0, \pm 1, \pm 2, \ldots, \pm\bar{c}. \quad (7.52)$$

Then, at each point of the first row of L_h, approximate $f(x,y)$ by means of

$$f(\bar{x} + mh, h) = \tfrac{1}{2}[p(\bar{x} + mh - h) + p(\bar{x} + mh + h)] + hq(\bar{x} + mh);$$
$$m = 0, \pm 1, \ldots, \pm(\bar{c} - 1). \quad (7.53)$$

Finally, in an inductive fashion, approximate $f(x,y)$ on the rth row of L_h, $r = 2,3, \ldots, \bar{c}$, in terms of its values already determined on the $(r - 1)$st and $(r - 2)$nd rows by means of relationship (4.5). In a finite number of steps, $f(\bar{x},\bar{y})$ is thus approximated.

Example. Let N be the set of real numbers. Let $u = f(x,y)$ be the solution on E^2 of the homogeneous wave equation (4.2) which satisfies the conditions

$$f(x,0) = 2x, \qquad\qquad x \in N$$

$$\frac{\partial f(x,0)}{\partial y} = \begin{cases} \sqrt{4 + x^3}, & x > 0 \\ 2, & x \leq 0. \end{cases}$$

Then approximate $f(3,3)$.

Solution. For $(\bar{x},\bar{y}) = (3,3)$, let $h = \tfrac{1}{2}$. Then, as shown in Diagram 7.6, the points of L_h, by rows, are: *row 0*: (0,0), (0.5,0), (1,0), (1.5,0), (2,0),

(2.5,0), (3,0), (3.5,0), (4,0), (4.5,0), (5,0), (5.5,0), (6,0); *row* 1: (0.5,0.5), (1,0.5), (1.5,0.5), (2,0.5), (2.5,0.5), (3,0.5), (3.5,0.5), (4,0.5), (4.5,0.5), (5,0.5), (5.5,0.5); *row* 2: (1,1), (1.5,1), (2,1), (2.5,1), (3,1), (3.5,1), (4,1), (4.5,1), (5,1); *row* 3: (1.5,1.5), (2,1.5), (2.5,1.5), (3,1.5), (3.5,1.5), (4,1.5), (4.5,1.5); *row* 4: (2,2), (2.5,2), (3,2), (3.5,2), (4,2); *row* 5: (2.5,2.5), (3,2.5), (3.5,2.5); *row* 6: (3,3).

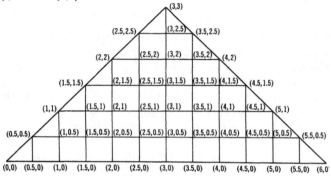

DIAGRAM 7.6

Since in the given example $p(x) \equiv 2x$, (7.52) implies that

$$f(0,0) = 0, \quad f(0.5,0) = 1, \quad f(1,0) = 2, \quad f(1.5,0) = 3,$$
$$f(2,0) = 4, \quad f(2.5,0) = 5, \quad f(3,0) = 6, \quad f(3.5,0) = 7,$$
$$f(4,0) = 8, \quad f(4.5,0) = 9, \quad f(5,0) = 10, \quad f(5.5,0) = 11,$$
$$f(6,0) = 12.$$

Application of (7.53), in order to approximate $f(x,y)$ on row 1, yields

$$f(0.5,0.5) = 0.5(\ 0 + \ 2) + 0.5\sqrt{4 + (0.5)^3} = \ 2.02$$
$$f(1,0.5) = 0.5(\ 1 + \ 3) + 0.5\sqrt{4 + 1^3} \quad = \ 3.12$$
$$f(1.5,0.5) = 0.5(\ 2 + \ 4) + 0.5\sqrt{4 + (1.5)^3} = \ 4.36$$
$$f(2,0.5) = 0.5(\ 3 + \ 5) + 0.5\sqrt{4 + 2^3} \quad = \ 5.73$$
$$f(2.5,0.5) = 0.5(\ 4 + \ 6) + 0.5\sqrt{4 + (2.5)^3} = \ 7.22$$
$$f(3,0.5) = 0.5(\ 5 + \ 7) + 0.5\sqrt{4 + 3^3} \quad = \ 8.78$$
$$f(3.5,0.5) = 0.5(\ 6 + \ 8) + 0.5\sqrt{4 + (3.5)^3} = 10.42$$
$$f(4,0.5) = 0.5(\ 7 + \ 9) + 0.5\sqrt{4 + 4^3} \quad = 12.12$$
$$f(4.5,0.5) = 0.5(\ 8 + 10) + 0.5\sqrt{4 + (4.5)^3} = 13.88$$
$$f(5,0.5) = 0.5(\ 9 + 11) + 0.5\sqrt{4 + 5^3} \quad = 15.68$$
$$f(5.5,0.5) = 0.5(10 + 12) + 0.5\sqrt{4 + (5.5)^3} = 17.53.$$

Use of (4.5) and the values of $f(x,y)$ already determined yields the following approximations for $f(x,y)$ on row 2:

$$
\begin{aligned}
f(1,1) &= f(0.5,0.5) + f(1.5,0.5) - f(1,0) &= 4.38 \\
f(1.5,1) &= f(1,0.5) \quad + f(2,0.5) \quad - f(1.5,0) &= 5.85 \\
f(2,1) &= f(1.5,0.5) + f(2.5,0.5) - f(2,0) &= 7.58 \\
f(2.5,1) &= f(2,0.5) \quad + f(3,0.5) \quad - f(2.5,0) &= 9.51 \\
f(3,1) &= f(2.5,0.5) + f(3.5,0.5) - f(3,0) &= 11.64 \\
f(3.5,1) &= f(3,0.5) \quad + f(4,0.5) \quad - f(3.5,0) &= 13.90 \\
f(4,1) &= f(3.5,0.5) + f(4.5,0.5) - f(4,0) &= 16.30 \\
f(4.5,1) &= f(4,0.5) \quad + f(5,0.5) \quad - f(4.5,0) &= 18.80 \\
f(5,1) &= f(4.5,0.5) + f(5.5,0.5) - f(5,0) &= 21.41.
\end{aligned}
$$

Use of (4.5) and the values of $f(x,y)$ already determined on rows 1 and 2 yields the following approximations for $f(x,y)$ on row 3:

$$
\begin{aligned}
f(1.5,1.5) &= f(1,1) \quad + f(2,1) \quad - f(1.5,0.5) &= 7.60 \\
f(2,1.5) &= f(1.5,1) + f(2.5,1) - f(2,0.5) &= 9.63 \\
f(2.5,1.5) &= f(2,1) \quad + f(3,1) \quad - f(2.5,0.5) &= 12.00 \\
f(3,1.5) &= f(2.5,1) + f(3.5,1) - f(3,0.5) &= 14.63 \\
f(3.5,1.5) &= f(3,1) \quad + f(4,1) \quad - f(3.5,0.5) &= 17.52 \\
f(4,1.5) &= f(3.5,1) + f(4.5,1) - f(4,0.5) &= 20.58 \\
f(4.5,1.5) &= f(4,1) \quad + f(5,1) \quad - f(4.5,0.5) &= 23.83.
\end{aligned}
$$

Use of (4.5) and the values of $f(x,y)$ already determined on rows 2 and 3 yields the following approximations for $f(x,y)$ on row 4:

$$
\begin{aligned}
f(2,2) &= f(1.5,1.5) + f(2.5,1.5) - f(2,1) &= 12.02 \\
f(2.5,2) &= f(2,1.5) \quad + f(3,1.5) \quad - f(2.5,1) &= 14.75 \\
f(3,2) &= f(2.5,1.5) + f(3.5,1.5) - f(3,1) &= 17.88 \\
f(3.5,2) &= f(3,1.5) \quad + f(4,1.5) \quad - f(3.5,1) &= 21.31 \\
f(4,2) &= f(3.5,1.5) + f(4.5,1.5) - f(4,1) &= 25.05.
\end{aligned}
$$

Use of (4.5) and the values of $f(x,y)$ already determined on rows 3 and 4 yields the following approximations for $f(x,y)$ on row 5:

$$
\begin{aligned}
f(2.5,2.5) &= f(2,2) \quad + f(3,2) \quad - f(2.5,1.5) &= 17.90 \\
f(3,2.5) &= f(2.5,2) + f(3.5,2) - f(3,1.5) &= 21.43 \\
f(3.5,2.5) &= f(3,2) \quad + f(4,2) \quad - f(3.5,1.5) &= 25.41.
\end{aligned}
$$

Finally, use of (4.5) and the values of $f(x,y)$ determined on rows 4 and 5 yields

$$f(3,3) = f(2.5,2.5) + f(3.5,2.5) - f(3,2) = 25.43 \sim 25.4.$$

Method III.   For Cauchy Problem II (Sec. 4.3), the numerical value of the solution $u = f(x,y)$ at any point $(\bar{x},\bar{y})$ can be calculated from (4.34).   If the integrals involved are not readily expressible in closed form, then, as in Method I, above, approximate methods of integration can be used.

Method IV.   For Cauchy Problem II, Method II above need only be modified by replacing (7.53) with

$$f(\bar{x} + mh, h) = \tfrac{1}{2}[p(\bar{x} + mh + h) + p(\bar{x} + mh - h)] + hq(\bar{x} + mh)$$

$$- \frac{h^2}{2} G(\bar{x} + mh, 0); \qquad m = 0, \pm 1, \ldots, \pm(\bar{c} - 1) \quad (7.54)$$

and (4.5) with

$$f(x + h, y) + f(x - h, y) - f(x, y + h) - f(x, y - h) = h^2 G(x,y). \tag{7.55}$$

Method V.   For the mixed-type problem (Sec. 4.4), if one seeks a solution $u = f(x,y)$ which is of class C^2 on $\bar{R}$, then the following method is available.   Let the points $(a,0)$, $(b,0)$ be designated P_1, P_2, respectively, as shown in Diagram 7.7.   Let the characteristic

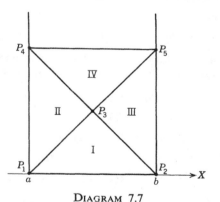

DIAGRAM 7.7

through P_1, with equation $y = x - a$, meet the line with equation $x = b$ in point P_5; let the characteristic through P_2, with equation $y = -x + b$, meet the line with equation $x = a$ in point P_4; and let the two characteristics under consideration meet in point P_3.   Then $a \leq x \leq b$ is the domain of dependence with respect to P_3.   Thus, at all points (x_1, y_1) either on or within the triangle $P_1 P_2 P_3$, labeled I in Diagram 7.7, the unique solution $u = f(x, y)$ can be determined either by the formula of D'Alembert or by Method I, above.

Next suppose that (x_1, y_1) is an arbitrary point on or within the triangle $P_1 P_3 P_4$, which is labeled II in Diagram 7.7. Let (x_1, y_1) then be denoted Q_1, as in Diagram 7.8. Let the characteristic

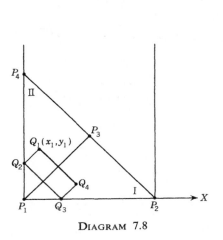

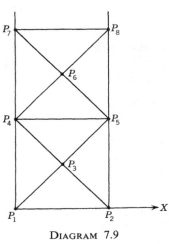

DIAGRAM 7.8   DIAGRAM 7.9

through Q_1 which has the equation $y - y_1 = x - x_1$ meet the line with equation $x = a$ in point Q_2. Using line segment $P_1 P_3$ as an axis of symmetry, construct rectangle $Q_1 Q_2 Q_3 Q_4$, as shown in Diagram 7.8. Since $f(Q_3)$, $f(Q_4)$ are known, because Q_3 lies on triangle I and Q_4 lies on or within triangle I, and since $f(Q_2)$ is known from (4.37), it follows that $f(Q_1)$ can be determined easily from (4.5). In this fashion then $f(x, y)$ can be determined at each point on or within triangle II.

In an analogous fashion, one can easily determine $f(x, y)$ at each point on or within the triangle $P_2 P_3 P_5$, labeled III in Diagram 7.7. Finally, using the known values of $f(x, y)$ on and within triangles I, II, and III, one can easily construct a rectangle, as indicated above, and use difference equation (4.5) to determine $f(x, y)$ at each point of the triangle $P_3 P_4 P_5$, labeled IV in Diagram 7.7.

Thus, in a constructive fashion, one can determine $f(x, y)$ on and within the square with vertices P_1, P_2, P_4, P_5, as shown in Diagram 7.7.

In the same fashion, $f(x, y)$ is readily determined on and within the square $P_4 P_5 P_7 P_8$, which has side $P_4 P_5$ in common with the square $P_1 P_2 P_4 P_5$ (see Diagram 7.9).

In an inductive fashion then, $u = f(x,y)$ is readily constructed and hence determined at each point of the semi-infinite strip $\bar{R}$, and the mixed-type problem is solved.

Example. For the mixed-type problem

$$u_{xx} - u_{yy} = 0$$
$$f(x,0) = \sin x, \qquad f_y(x,0) = 1, \qquad 0 \le x \le \pi$$
$$f(0,y) = f(\pi,y) = 0, \qquad y \ge 0,$$

assume there exists a unique solution of class C^2 and find $f\left(\dfrac{\pi}{4}, \dfrac{\pi}{4}\right)$, $f\left(\dfrac{\pi}{2}, \dfrac{\pi}{2}\right)$, $f\left(\dfrac{\pi}{4}, \dfrac{3\pi}{4}\right)$, $f\left(\dfrac{3\pi}{4}, \dfrac{5\pi}{4}\right)$.

Solution. Noting that the points $\left(\dfrac{\pi}{4}, \dfrac{\pi}{4}\right)$, $\left(\dfrac{\pi}{2}, \dfrac{\pi}{2}\right)$, $\left(\dfrac{\pi}{4}, \dfrac{3\pi}{4}\right)$, $\left(\dfrac{3\pi}{4}, \dfrac{5\pi}{4}\right)$

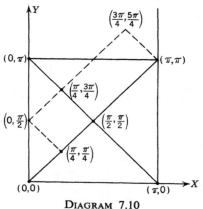

DIAGRAM 7.10

are arranged as shown in Diagram 7.10, then Method I, above, yields

$$f\left(\frac{\pi}{4}, \frac{\pi}{4}\right) \sim 1.29$$

$$f\left(\frac{\pi}{2}, \frac{\pi}{2}\right) \sim 1.57.$$

Since $m_1(y) = m_2(y) = 0$, $y \ge 0$, it follows from (4.5) and the results just established that

$$f\left(\frac{\pi}{4}, \frac{3\pi}{4}\right) = f\left(0, \frac{\pi}{2}\right) + f\left(\frac{\pi}{2}, \frac{\pi}{2}\right) - f\left(\frac{\pi}{4}, \frac{\pi}{4}\right) \sim 0.28,$$

$$f\left(\frac{3\pi}{4}, \frac{5\pi}{4}\right) = f(\pi,\pi) + f\left(0, \frac{\pi}{2}\right) - f\left(\frac{\pi}{4}, \frac{\pi}{4}\right) \sim -1.29.$$

Finally, note that in this example the D'Alembert formula could also have been used easily to yield the exact values

$$f\left(\frac{\pi}{4}, \frac{\pi}{4}\right) = \frac{1}{2} + \frac{\pi}{4},$$

$$f\left(\frac{\pi}{2}, \frac{\pi}{2}\right) = \frac{\pi}{2},$$

and thus (4.5), as applied above, would also yield the exact values

$$f\left(\frac{\pi}{4}, \frac{3\pi}{4}\right) = \frac{\pi}{4} - \frac{1}{2},$$

$$f\left(\frac{3\pi}{4}, \frac{5\pi}{4}\right) = -\frac{\pi}{4} - \frac{1}{2}.$$

Method VI (Crank-Nicolson Method). For the boundary-value problem of Type H (Sec. 6.1), divide $0 \leq x \leq \pi$ into r_1 equal parts such that $\pi/r_1 = h_1$ and divide $0 \leq y \leq b$ into r_2 equal parts such that $b/r_2 = h_2$. If $u = f(x,y)$ is the analytical solution of the problem, then by (6.2) to (6.4), $f(x,y)$ is known on $L_1 \cup L_2 \cup L_3$. The technique to be described, then, will indicate how to approximate $f(x,y)$ at various points of $R \cup L_4$.

At each of the points (h_1, h_2), $(2h_1, h_2)$, $\ldots$, $(r_1 h_1 - h_1, h_2)$ approximate $f(x,y)$ by applying the difference equation

$$\frac{f(x,y) - f(x, y - h_2)}{h_2} = \frac{1}{2}\left[\frac{f(x + h_1, y) - 2f(x,y) + f(x - h_1, y)}{h_1^2}\right.$$

$$\left. + \frac{f(x + h_1, y - h_2) - 2f(x, y - h_2) + f(x - h_1, y - h_2)}{h_1^2}\right] \quad (7.56)$$

and by solving the resulting system of $(r_1 - 1)$ equations in the $(r_1 - 1)$ unknown function values.

Next, at each of the points $(h_1, 2h_2)$, $(2h_1, 2h_2)$, $(3h_1, 2h_2)$, $\ldots$ $(r_1 h_1 - h_1, 2h_2)$ approximate $f(x,y)$ by applying the difference equation (7.56). Using the known values of $f(x,y)$ on $L_1 \cup L_3$ and the approximate values found above, this process yields a system of $(r_1 - 1)$ equations in $(r_1 - 1)$ unknown function values whose solution constitutes the desired approximation.

Continue then in an inductive fashion to apply (7.56) at each of the points (h_1, nh_2), $(2h_1, nh_2)$, $(3h_1, nh_2)$, $\ldots$, $(r_1 h_1 - h_1, nh_2)$. Using the values of $f(x,y)$ on $L_1 \cup L_3$ and the set of approximations at $(h_1, (n - 1)h_2)$, $(2h_1, (n - 1)h_2)$, $\ldots$, $(r_1 h_1 - h_1, (n - 1)h_2)$ yields a system of $(r_1 - 1)$ equations in $(r_1 - 1)$ function values whose solution constitutes the desired approximation. The final approximations result when $n = r_2$ and the process terminates.

Example. For the boundary-value problem of type H with

$$f(0,y) \equiv 0 \qquad \text{on } L_1 \qquad\qquad (7.57)$$

$$f(x,0) \equiv \sin x \qquad \text{on } L_2 \qquad\qquad (7.58)$$

$$f(\pi,y) \equiv 0 \qquad \text{on } L_3, \qquad\qquad (7.59)$$

find an approximate value of the solution $u = f(x,y)$ at $\left(\dfrac{\pi}{3}, \dfrac{\pi^2}{9}\right)$.

Solution. Allowing $r_1 = 3$, $r_2 = 2$, $b = \pi^2/9$ yields $h_1 = \pi/3$, $h_2 = \pi^2/18$. The points which must then be considered in applying the Crank-Nicolson

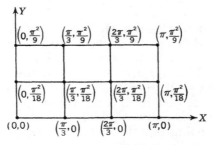

DIAGRAM 7.11

method are labeled in Diagram 7.11. Application of (7.56) at the points $\left(\dfrac{\pi}{3}, \dfrac{\pi^2}{18}\right)$, $\left(\dfrac{2\pi}{3}, \dfrac{\pi^2}{18}\right)$ and use of (7.57) to (7.59) yield the system of equations

$$\frac{f\left(\dfrac{\pi}{3}, \dfrac{\pi^2}{18}\right) - \dfrac{\sqrt{3}}{2}}{\dfrac{\pi^2}{18}} = \frac{1}{2}\left[\frac{f\left(\dfrac{2\pi}{3}, \dfrac{\pi^2}{18}\right) - 2f\left(\dfrac{\pi}{3}, \dfrac{\pi^2}{18}\right)}{\dfrac{\pi^2}{9}} + \frac{\dfrac{\sqrt{3}}{2} - \dfrac{2\sqrt{3}}{2} + 0}{\dfrac{\pi^2}{9}}\right]$$

$$\frac{f\left(\dfrac{2\pi}{3}, \dfrac{\pi^2}{18}\right) - \dfrac{\sqrt{3}}{2}}{\dfrac{\pi^2}{18}} = \frac{1}{2}\left[\frac{0 - 2f\left(\dfrac{2\pi}{3}, \dfrac{\pi^2}{18}\right) + f\left(\dfrac{\pi}{3}, \dfrac{\pi^2}{18}\right)}{\dfrac{\pi^2}{9}} + \frac{0 - \dfrac{2\sqrt{3}}{2} + \dfrac{\sqrt{3}}{2}}{\dfrac{\pi^2}{9}}\right],$$

the solution of which is

$$f\left(\frac{\pi}{3}, \frac{\pi^2}{18}\right) = f\left(\frac{2\pi}{3}, \frac{\pi^2}{18}\right) = \frac{3\sqrt{3}}{10}. \qquad\qquad (7.60)$$

Application of (7.56) at the points $\left(\dfrac{\pi}{3}, \dfrac{\pi^2}{9}\right)$, $\left(\dfrac{2\pi}{3}, \dfrac{\pi^2}{9}\right)$ and use of (7.57), (7.59), and (7.60) yield the system of equations

$$\dfrac{f\left(\dfrac{\pi}{3}, \dfrac{\pi^2}{9}\right) - \dfrac{3\sqrt{3}}{10}}{\dfrac{\pi^2}{18}} = \dfrac{1}{2}\left[\dfrac{f\left(\dfrac{2\pi}{3}, \dfrac{\pi^2}{9}\right) - 2f\left(\dfrac{\pi}{3}, \dfrac{\pi^2}{9}\right) + 0}{\dfrac{\pi^2}{9}} + \dfrac{\dfrac{3\sqrt{3}}{10} - \dfrac{3\sqrt{3}}{5} + 0}{\dfrac{\pi^2}{9}}\right]$$

$$\dfrac{f\left(\dfrac{2\pi}{3}, \dfrac{\pi^2}{9}\right) - \dfrac{3\sqrt{3}}{10}}{\dfrac{\pi^2}{18}} = \dfrac{1}{2}\left[\dfrac{0 - 2f\left(\dfrac{2\pi}{3}, \dfrac{\pi^2}{9}\right) + f\left(\dfrac{\pi}{3}, \dfrac{\pi^2}{9}\right)}{\dfrac{\pi^2}{9}} + \dfrac{0 - \dfrac{3\sqrt{3}}{5} + \dfrac{3\sqrt{3}}{10}}{\dfrac{\pi^2}{9}}\right],$$

the solution of which is

$$f\left(\frac{\pi}{3}, \frac{\pi^2}{9}\right) = f\left(\frac{2\pi}{3}, \frac{\pi^2}{9}\right) = \frac{9\sqrt{3}}{50}, \tag{7.61}$$

so that, finally,

$$f\left(\frac{\pi}{3}, \frac{\pi^2}{9}\right) \sim 0.31.$$

Method VII. For the boundary-value problem of type H, proceed as in Method VI, above, but select $h_2/(h_1^2) \le \frac{1}{2}$ and replace (7.56) with

$$\frac{f(x,y) - f(x, y - h_2)}{h_2} =$$

$$\frac{f(x + h_1, y - h_2) - 2f(x, y - h_2) + f(x - h_1, y - h_2)}{h_1^2} \tag{7.62}$$

Finally, it should be noted that the techniques described in this chapter are particularly amenable to high-speed digital computation, that approximation methods associated with the heat equation have a special pathology, and that a large variety of other methods are available.

EXERCISES

7.1. For $(\bar{x}, \bar{y}) = (0,0)$, $h = 1$, find approximate solutions for each of the following Dirichlet problems by means of the Liebmann-Gerschgorin-Collatz procedure:

(a) Let B be the triangle with vertices $(-5,0)$, $(5,0)$, $(0,4)$, and let R be its interior. Let $g(x,y) = 25 - x^2$ on the side of B which connects $(-5,0)$ and $(5,0)$, and let $g(x,y) = 0$ on the remainder of B.

(b) Let B be the rectangle with vertices $(-4,0)$, $(4,0)$, $(4,3)$, $(-4,3)$, and let R be its interior.   Let $g(x,y) = 16 - x^2$ on the side B which connects $(-4,0)$ and $(4,0)$, and let $g(x,y) = 0$ on the remainder of B.

(c) Let B be the circle with center $(0,0)$, radius 4 and let R be its interior.   Let $g(x,y) = 1 - x^2 + y$ on B.

(d) Let B be the ellipse whose equation is $\dfrac{x^2}{16} + \dfrac{y^2}{9} = 1$, and let $g(x,y) = 1 - x^2 + y$ on B.

(e) Let B be the hexagon whose consecutive vertices are $(-4,0)$, $(4,0)$, $(4,5)$, $(0,5)$, $(0,4)$, $(-4,4)$, and let R be its interior.   Let $g(x,y) = 16 - x^2$ on the side of B which connects $(-4,0)$ and $(4,0)$, and let $g(x,y) = 0$ on the remainder of B.

7.2. For $(\bar{x},\bar{y}) = (0,0)$, $h = 0.5$, find approximate solutions for each of the Dirichlet problems (a) to (c) of Exercise 7.1.

7.3. Let $u = f(x,y)$ represent the analytical solution of Cauchy Problem I. Then, by Method I, approximate $f(x^*,y^*)$ in each of the following cases:

(a) $p(x) = x$, $q(x) = \sqrt{1 + x^4}$, $(x^*,y^*) = (3,3)$

(b) $p(x) = 1$, $q(x) = \sqrt{5 + x^4}$, $(x^*,y^*) = (3,3)$

(c) $p(x) = x^3$, $q(x) = \sqrt[3]{1 + x^2}$, $(x^*,y^*) = (3,3)$

(d) $p(x) = x$, $q(x) = \sqrt{1 + x^4}$, $(x^*,y^*) = (1,4)$

(e) $p(x) = 1$, $q(x) = \sqrt{5 + x^4}$, $(x^*,y^*) = (-1,4)$

(f) $p(x) = x^3$, $q(x) = \sqrt[3]{1 + x^2}$, $(x^*,y^*) = (-1,-4)$

(g) $p(x) = x^2 - 1$, $q(x) = \dfrac{e^x}{1 + x^2}$, $(x^*,y^*) = (3,3)$

(h) $p(x) = e^x$, $q(x) = \dfrac{\sin x}{1 + x^4}$, $(x^*,y^*) = (0,4)$.

7.4. Find approximations for each of the Cauchy problems in Exercise 7.3, above, by means of Method II.

7.5. Let $u = f(x,y)$ represent the analytical solution of Cauchy Problem II. Then, by Method III, approximate $f(x^*,y^*)$ in each of the following cases:

(a) $G(x,y) = 1$, $p(x) = x$, $q(x) = \sqrt{1 + x^4}$, $(x^*,y^*) = (3,3)$

(b) $G(x,y) = x$, $p(x) = x$, $q(x) = \sqrt{1 + x^4}$, $(x^*,y^*) = (3,3)$

(c) $G(x,y) = y^2$, $p(x) = x$, $q(x) = \sqrt{1 + x^4}$, $(x^*,y^*) = (3,3)$

(d) $G(x,y) = xy^2$, $p(x) = x$, $q(x) = \sqrt{1 + x^4}$, $(x^*,y^*) = (3,3)$

(e) $G(x,y) = xy^3$, $p(x) = 1 - x^2$, $q(x) = \dfrac{\sin x}{1 + x^4}$, $(x^*,y^*) = (0,4)$.

7.6. Find approximations for each of the Cauchy problems in Exercise 7.5, above, by means of Method IV.

7.7. For the mixed-type problem, assume there exists a solution $u = f(x,y)$ of class C^2. Then in each of the following cases, approximate $f\left(\dfrac{\pi}{4},\dfrac{\pi}{4}\right)$, $f\left(\dfrac{\pi}{2},\dfrac{\pi}{2}\right)$, $f\left(\dfrac{\pi}{4},\dfrac{3\pi}{4}\right)$, $f\left(\dfrac{3\pi}{4},\dfrac{5\pi}{4}\right)$, $f\left(\dfrac{3\pi}{4},2\pi\right)$, $f\left(\dfrac{\pi}{4},\dfrac{9\pi}{4}\right)$ by Method V:

(a) $f(x,0) = \sin x$, $f_y(x,0) = 2$, $0 \leq x \leq \pi$
$f(0,y) = f(\pi,y) = 0$, $y \geq 0$

(b) $f(x,0) = \sin x,$ $\qquad f_y(x,0) = 1 - x,$ $\qquad 0 \le x \le \pi$
$\quad f(0,y) = f(\pi,y) = 0,$ $\qquad\qquad\qquad\qquad y \ge 0$

(c) $f(x,0) = \sin x,$ $\qquad f_y(x,0) = x^3,$ $\qquad 0 \le x \le \pi$
$\quad f(0,y) = f(\pi,y) = 0,$ $\qquad\qquad\qquad\qquad y \ge 0$

(d) $f(x,0) = \cos x,$ $\qquad f_y(x,0) = 1 + x,$ $\qquad 0 \le x \le \pi$
$\quad f(0,y) = 1,$ $\qquad\qquad f(\pi,y) = -1,$ $\qquad y \ge 0$

(e) $f(x,0) = \sin x,$ $\qquad f_y(x,0) = e^x,$ $\qquad 0 \le x \le \pi$
$\quad f(0,y) = \sin y,$ $\qquad f(\pi,y) = y^2 - y,$ $\qquad y \ge 0.$

7.8. Let $u = f(x,y)$ represent the analytical solution of the boundary-value problem of type H. Then, by Method VI, approximate $f(x^*,y^*)$ in each of the following cases:

(a) $f(0,y) = 0$ on $L_1,$ $f(x,0) = \sin 2x$ on $L_2,$ $f(\pi,y) = 0$ on $L_3,$
$\quad (x^*,y^*) = \left(\dfrac{\pi}{3}, \dfrac{\pi^2}{9}\right),$ $\quad r_1 = 3,$ $\quad r_2 = 2,$ $\quad b = \dfrac{\pi^2}{9}.$ (*Ans.* Approx. 0.18)

(b) $f(0,y) = 0$ on $L_1,$ $f(x,0) = \sin x + \sin 2x$ on $L_2,$ $f(\pi,y) = 0$ on $L_3,$
$\quad (x^*,y^*) = \left(\dfrac{2\pi}{3}, \dfrac{\pi^2}{9}\right),$ $\quad r_1 = 3,$ $\quad r_2 = 2,$ $\quad b = \dfrac{\pi^2}{9}.$ (*Ans.* Approx. 0.29)

(c) $f(0,y) = 0$ on $L_1,$ $f(x,0) = \sin x$ on $L_2,$ $f(\pi,y) = 0$ on $L_3,$
$\quad (x^*,y^*) = \left(\dfrac{\pi}{3}, \dfrac{\pi^2}{9}\right),$ $\quad r_1 = 6,$ $\quad r_2 = 2,$ $\quad b = \dfrac{\pi^2}{9}.$

(d) $f(0,y) = 0$ on $L_1,$ $f(x,0) = \sin x$ on $L_2,$ $f(\pi,y) = 0$ on $L_3,$
$\quad (x^*,y^*) = \left(\dfrac{\pi}{3}, \dfrac{\pi^2}{9}\right),$ $\quad r_1 = 3,$ $\quad r_2 = 4,$ $\quad b = \dfrac{\pi^2}{9}.$

(e) $f(0,y) = 0$ on $L_1,$ $f(x,0) = \sin x$ on $L_2,$ $f(\pi,y) = 0$ on $L_3,$
$\quad (x^*,y^*) = \left(\dfrac{\pi}{3}, \dfrac{\pi^2}{9}\right),$ $\quad r_1 = 6,$ $\quad r_2 = 4,$ $\quad b = \dfrac{\pi^2}{9}.$

7.9. Find approximations for each of the boundary-value problems of type H in Exercise 7.8 above, by means of Method VII.

Chapter 8

SURVEY OF OTHER TOPICS

The field of partial differential equations is resplendent with topics for further study and a few of those which have almost universal appeal will be described now.   The discussions, of necessity, will be far more intuitive than rigorous.

8.1   Other Methods of Solution

Besides the method of separation of variables, which depends extensively on the theory of Fourier series, there are only a few other *general* methods available for solving large classes of partial differential equations.   One such technique is the method of transforms and a variety of supporting theories are available depending on whether the type of transform used is the Laplace transform, the Fourier transform, or, say, the Mellin transform [28], [35], [36], [40], [44], [47].

Without proof, some salient features of the theory and application of Laplace transforms are now presented.

Definition 8.1.   Let N be the set of real numbers.   If $x, y, t \in N$, $z = x + iy$, and $f(t)$ is a real-valued function of t, then, provided all integrals under consideration exist, the Laplace transform $F(z)$ of $f(t)$ is defined by

$$F(z) \equiv L[f(t)] \equiv \int_0^\infty e^{-zt} f(t) \, dt. \tag{8.1}$$

It can be shown that each of the following is a valid example of a Laplace transform.

Example 1.

$$L[1] = \frac{1}{z}, \qquad x > 0.$$

Example 2.

$$L[e^{\alpha t}] = \frac{1}{z - \alpha}, \qquad \alpha = a + ib, a \in N, b \in N, x > a.$$

Example 3.

$$L[\sin \alpha t] = \frac{\alpha}{z^2 + \alpha^2}, \qquad \alpha = a + ib, a \in N, b \in N, x > 0.$$

Example 4.

$$L[\cos \alpha t] = \frac{z}{z^2 + \alpha^2}, \qquad \alpha = a + ib, a \in N, b \in N, x > 0.$$

Example 5.

$$L[t^{n-1}] = \frac{(n-1)!}{z^n}, \qquad x > 0, n = 1, 2, 3, \ldots$$

Example 6. If $J_n(t)$ is what is known as the Bessel function of order n ([14], [44]), then

$$L[J_n(t)] = \left(\frac{\sqrt{z^2 + 1} - z}{\sqrt{z^2 + 1}} \right)^n, \qquad z \neq \pm i.$$

A fundamental theorem relating the Laplace transform of a function to that of its derivative is stated as follows. The proof requires only integration by parts, as applied to complex-valued functions.

Theorem 8.1. If $L[f(t)]$, $L[f'(t)]$ exist and

$$\lim_{t \to \infty} e^{-zt} f(t) = 0, \tag{8.2}$$

then

$$L[f'(t)] = zL[f(t)] - f(0). \tag{8.3}$$

Under conditions analogous to those stated in Theorem 8.1, it readily follows that

$$L[f''(z)] = z^2 L[f(t)] - zf(0) - f'(0), \tag{8.4}$$

$$L[f'''(z)] = z^3 L[f(t)] - z^2 f(0) - zf'(0) - f''(0), \tag{8.5}$$

and so forth.

Now, with just the few ideas, examples, and results described above, one can show how the concept of a transform applies to the solution of partial differential equations.

Example. Find a solution $u = f(x,y)$ of the homogeneous wave equation

$$\frac{\partial^2 u}{\partial x^2} - \frac{\partial^2 u}{\partial y^2} = 0 \tag{8.6}$$

which satisfies the conditions

$$f(x,0) = 0, \qquad x \geq 0 \tag{8.7}$$

$$\frac{\partial f(x,0)}{\partial y} = 0, \qquad x \geq 0 \tag{8.8}$$

$$f(0,y) = y^2, \qquad y \geq 0 \tag{8.9}$$

$$\lim_{x \to \infty} f(x,y) = 0, \qquad y \geq 0. \tag{8.10}$$

Solution. In order not to introduce notational difficulties, it will be convenient in the discussion to view x, y, z as three independent variables. Moreover, the discussion will use desirable, but not necessarily valid, steps, so that it will be necessary to verify the final result.

Multiplying, then, both sides of (8.6) by e^{-zy} and integrating, as indicated, yields

$$\int_0^\infty \frac{\partial^2 u(x,y)}{\partial x^2} e^{-zy} \, dy - \int_0^\infty \frac{\partial^2 u(x,y)}{\partial y^2} e^{-zy} \, dy = 0. \tag{8.11}$$

Assuming

$$\int_0^\infty \frac{\partial^2 u(x,y)}{\partial x^2} e^{-zy} \, dy \equiv \frac{\partial^2}{\partial x^2} \int_0^\infty u(x,y) e^{-zy} \, dy, \tag{8.12}$$

and since, by (8.4),

$$\int_0^\infty \frac{\partial^2 u(x,y)}{\partial y^2} e^{-zy} \, dy = z^2 L[u(x,y)] - zu(x,0) - \frac{\partial u(x,0)}{\partial y}, \tag{8.13}$$

then, (8.11) to (8.13) imply

$$\frac{\partial^2}{\partial x^2} \int_0^\infty u(x,y) e^{-zy} \, dy = z^2 L[u(x,y)] - zu(x,0) - \frac{\partial u(x,0)}{\partial y} \tag{8.14}$$

By (8.1), (8.7), and (8.8), (8.14) reduces to

$$\frac{\partial^2}{\partial x^2} L[u(x,y)] - z^2 L[u(x,y)] = 0. \tag{8.15}$$

A solution of (8.15) is

$$L[u(x,y)] = A(z)e^{zx} + B(z)e^{-zx}. \tag{8.16}$$

In order to satisfy (8.10), allow $A(z) \equiv 0$. In order to satisfy (8. 9), allow

$$B(z) = \int_0^\infty u(0,y) e^{-zy} \, dy = \int_0^\infty y^2 e^{-zy} \, dy. \tag{8.17}$$

Then (8.1), (8.16), and (8.17) imply

$$\int_0^\infty u(x,y)e^{-zy}\,dy = \int_0^\infty y^2 e^{-zy}e^{-zx}\,dy,$$

or, equivalently, that

$$\int_0^\infty u(x,y)e^{-zy}\,dy = \int_0^\infty t^2 e^{-zt}e^{-zx}\,dt. \tag{8.18}$$

Under the transformation $y = x + t$, (8.18) can be rewritten

$$\int_0^\infty u(x,y)e^{-zy}\,dy = \int_x^\infty (y - x)^2 e^{-zy}\,dy. \tag{8.19}$$

Finally, setting

$$(y - x) \equiv 0, \qquad 0 \le y \le x \tag{8.20}$$

enables one to write (8.19) as

$$\int_0^\infty u(x,y)e^{-zy}\,dy = \int_0^\infty (y - x)^2 e^{-zy}\,dy. \tag{8.21}$$

Thus, (8.20) and (8.21) suggest

$$f(x,y) = \begin{cases} (y - x)^2, & y > x \ge 0 \\ 0, & 0 \le y \le x \end{cases} \tag{8.22}$$

as a solution to the problem. Simple calculation reveals that (8.22) is such a solution.

Another method of great value, which, due to its dependence on the theory of analytic continuation and Fourier transforms, is only mentioned by name, is the Wiener-Hopf technique [36].

A final, general method worthy of note is the method of Green's functions, which is particularly useful in the study of elliptic partial differential equations.

Definition 8.2. Let R be a simply connected, bounded region whose boundary B is a contour. The Green's function $\mathcal{G}(x,y;\xi,\eta)$ for $R \cup B$ is a function of the two points (x,y), (ξ,η), where $(x,y) \in R$, $(\xi,\eta) \in (R \cup B)$, with the following properties:

(a) for (x,y) fixed, $\mathcal{G}$ is a regular, harmonic function in the variables ξ, η on R, *except* at (x,y), and is continuous on $R \cup B$,

(b) $\mathcal{G}$ becomes logarithmically infinite as $(\xi,\eta) \to (x,y)$ in such a way that

$$\mathcal{G}(x,y;\xi,\eta) - \log \frac{1}{\sqrt{(x - \xi)^2 + (y - \eta)^2}} \equiv W(x,y;\xi,\eta)$$

is a regular, harmonic function at (x,y),

(c) if $(x,y) \in R$, $(\xi,\eta) \in B$, then $\mathcal{G}(x,y;\xi,\eta) = 0$.

Example. If B is the unit circle and R is its interior, then

$$\mathscr{G}(x,y;\xi,\eta) = \log \sqrt{\frac{1 - 2(x\xi + y\eta) + (x^2 + y^2)(\xi^2 + \eta^2)}{(x - \xi)^2 + (y - \eta)^2}}. \quad (8.23)$$

By means of the Green's function, one can give (see Ref. 46) the solution $u = f(x,y)$ of the Dirichlet problem in terms of the integral

$$f = -\frac{1}{2\pi} \oint_B g \frac{\partial \mathscr{G}}{\partial n} \, ds. \quad (8.24)$$

Thus, (8.23) and (8.24) can be combined readily to yield the Poisson integral formula. However, though the existence of the Green's function can be established [47], producing it in the case where B is not necessarily circular is a problem of great difficulty.

8.2   The Cauchy-Kowalewski Theorem

Much energy has been and is being directed toward the determination of whether or not a given partial differential equation, or system of such equations, has a solution. The analogy between the concept of a fixed point in functional analysis and that of a solution of a differential equation has yielded many results in this direction. However, the classical result of this field of study is the Cauchy-Kowalewski theorem, which will now be stated for a system in which there are two independent variables. For the most general form of the theorem consult Ref. 37.

Theorem 8.2.   Cauchy-Kowalewski Theorem. Let x, y be independent variables and let $u_1, u_2, \ldots, u_N$ be dependent variables. Consider the system of partial differential equations

$$\frac{\partial^{n_i} u_i}{\partial x^{n_i}} = F_i\left(x, y, u_1, u_2, \ldots, u_N, \ldots, \frac{\partial^k u_j}{\partial x^{k_0} \partial y^{k_1}}, \ldots\right), \quad (8.25)$$

where $i, j = 1, 2, \ldots, N$; $k_0 + k_1 = k \leq n_j$, $k_0 < n_j$. For some value $x = x_0$, prescribe function values and derivatives up to and including order $n_i - 1$ by means of

$$\frac{\partial^k u_i}{\partial x^k} = \phi_i^{(k)}(y); \qquad k = 0, 1, 2, \ldots, n_i - 1, \quad (8.26)$$

where the derivative of order zero is defined to be the function itself. With respect to (8.26), we shall use the following notation for the derivatives of the functions $\phi_i^{(k)}$ at the fixed value $y = y_0$:

$$\left.\frac{\partial^{k_1} \phi_i^{(k_0)}}{\partial y^{k_1}}\right|_{y=y_0} = \phi_{i,k_0,k_1}^0, \quad (8.27)$$

where $i = 1, 2, \ldots, N$; $k_0 + k_1 = k \le n_i$. If all the functions F_i are analytic in some spherical neighborhood of the point $(x_0, y_0, \ldots, \phi_{j,k_0,k_1}^0, \ldots)$, and if all the functions $\phi_i^{(k)}(y)$ are analytic in some neighborhood of $y = y_0$, then there exists a unique set of functions $u_i = f_i(x, y)$, $i = 1, 2, \ldots, N$, each of which is analytic and satisfies (8.25) and (8.26) in some neighborhood of (x_0, y_0).

8.3 Remarks about Nonlinear Differential Equations

The most elusive and least understood aspect of partial differential equations is the study of nonlinear equations. The entire field consists essentially of a few isolated methods which apply to only a few ultraspecial equations (see, for example, Bers[4], Greenspan and Carrier [16], and Lighthill [32]). In the physical sciences, where the development of such precise instruments for measurement as the electron microscope and the ammonia clock has induced extensive study of nonlinear models, the absence of mathematical method has caused and will continue to cause great hardship. The rates of growth of these sciences are impeded by the necessity to resort at each developmental stage to extensive, time-consuming experimentation at a time when mathematical methods should be available. Much mathematical research and study must therefore be supported and encouraged in this direction.

SELECTED BIBLIOGRAPHY

1. P. S. Aleksandrov: "Combinatorial Topology," Rochester, N.Y., Graylock, 1956.
2. H. Bateman: "Partial Differential Equations of Mathematical Physics," Cambridge, Cambridge University Press, 1932.
3. D. L. Bernstein: "Existence Theorems in Partial Differential Equations," Princeton, N.J., Princeton University Press, 1950.
4. L. Bers: On Mildly Nonlinear Partial Differential Equations of Elliptic Type, *Jour. Research Natl. Bur. Standards*, **51**: 229–236 (1953).
5. R. V. Churchill: "Fourier Series and Boundary Value Problems," New York, McGraw-Hill, 1941.
6. L. Collatz: "Numerische Behandlung von Differentialgleichungen," Berlin, Springer, 1951.
7. R. Courant: "Differential and Integral Calculus," vol. I, New York, Interscience, 1947; vol. II, London, Blackie, 1940.
8. R. Courant, K. Friedrichs, and H. Lewy: Uber die partiellen Differentialgleichungen der mathematischen Physik, *Math. Annalen*, **100**: 32–74 (1928).
9. R. Courant and D. Hilbert: "Methoden der mathematischen Physik," vol II, Berlin, Springer, 1931.
10. J. Crank and P. Nicolson: A Practical Method for Numerical Evaluation of Solutions of Partial Differential Equations of Heat-conduction Type, *Proc. Camb. Philos. Soc.*, **43**: 50–67 (1947).
11. G. F. D. Duff: "Partial Differential Equations," Toronto, University of Toronto Press, 1956.
12. G. Fichera: On a Unified Theory of Boundary Value Problems for Elliptic-parabolic Equations of Second Order, *MRC Technical Summary Report* 110, Math. Res. Center, U.S. Army, University of Wisconsin, 1959.
13. K. O. Friedrichs: Symmetric Positive Linear Differential Equations, *Comm. Pure Appl. Math.*, **11**: 333–418 (1958).
14. D. Greenspan: "Theory and Solution of Ordinary Differential Equations," New York, Macmillan, 1960.
15. D. Greenspan: On the Numerical Solution of Dirichlet-type Problems, *Amer. Math. Mo.*, **LXVI**: 40–46 (1959).
16. H. P. Greenspan and G. F. Carrier: The Magnetohydrodynamic Flow Past a Flat Plate, *Journ. Fluid. Mech.*, **6**: 77–96 (1959).

17. J. S. Hadamard: "Lectures on Cauchy's Problem in Linear Partial Differential Equations," New York, Dover, 1952.
18. D. W. Hall and G. L. Spencer II: "Elementary Topology," New York, Wiley, 1955.
19. H. S. Hall and S. R. Knight, "Higher Algebra," London, Macmillan, 1957.
20. D. R. Hartree: "Numerical Analysis," Oxford, Clarendon Press, 1952.
21. F. B. Hildebrand: "Advanced Calculus for Engineers," Englewood Cliffs, N.J., Prentice-Hall, 1949.
22. J. Horn: "Einfuhrung in die Theorie der partiellen Differentialgleichungen," Leipzig, Goshen, 1910.
23. D. Jackson: "Fourier Series and Orthogonal Polynomials," Menasha (Wisconsin), Mathematical Association of America, 1941.
24. R. L. Jeffrey: "Trigonometric Series," Toronto, University of Toronto Press, 1956.
25. F. John: "Partial Differential Equations," lecture notes, New York University, 1952–1953.
26. F. John: "Advanced Numerical Analysis," lecture notes, New York University, 1956.
27. L. V. Kantorovich and V. I. Krylov: "Approximate Methods of Higher Analysis," New York, Interscience, 1958.
28. H. Lass: "Elements of Pure and Applied Mathematics," New York, McGraw-Hill, 1957.
29. P. Lax: "Partial Differential Equations," lecture notes, New York University, 1949–1950.
30. J. Leray: "Hyperbolic Differential Equations," lecture notes, Princeton, N.J., Institute for Advanced Study, 1955.
31. H. Liebmann: Die angenahrte Ermittelung harmonischer Funktionen und konformer Abbildungen, *Sitz. Bayer Akad. Wiss. Math-Phys. Klasse,* pp. 385–416, 1918.
32. M. J. Lighthill: A Technique for Rendering Approximate Solutions to Physical Problems Uniformly Valid, *Phil. Mag.,* **40:** 1179–1202 (1949).
33. K. S. Miller: "Partial Differential Equations in Engineering Problems," Englewood Cliffs, N.J., Prentice-Hall, 1953.
34. W. E. Milne: "Numerical Solution of Differential Equations," New York, Wiley, 1953.
35. P. M. Morse and H. Feshbach: "Methods of Theoretical Physics," vols. I, II, New York, McGraw-Hill, 1953.
36. B. Noble: "Methods Based on the Weiner-Hopf Technique for the Solution of Partial Differential Equations," New York, Pergamon, 1958.
37. I. G. Petrovsky: "Lectures on Partial Differential Equations," New York, Interscience, 1954.
38. L. A. Pipes: "Applied Mathematics for Engineers and Physicists," 2d ed., York, McGraw-Hill, 1958.
39. R. D. Richtmyer: "Difference Methods for Initial Value Problems," New York, Interscience, 1957.
40. P. C. Rosenbloom: "Linear Partial Differential Equations," lecture notes, Harvard University, 1957.
41. M. G. Salvadori: "Differential Equations in Engineering Problems," Englewood Cliffs, N.J., Prentice-Hall, 1954.
42. R. Sauer: "Anfangswertprobleme bei partiellen Differentialgleichungen," Berlin, Springer, 1952.

43. I. N. Sneddon, "Elements of Partial Differential Equations," New York, McGraw-Hill, 1957.

44. S. L. Sobolew: Einige Anwendungen der Funktionalanalysis in der mathematischen Physik, Leningrad, Izdat. Leningrad Gos. Univ., 1950.

45. A. J. W. Sommerfeld: "Partial Differential Equations in Physics," New York, Academic Press, 1949.

46. W. J. Sternberg and T. L. Smith: "The Theory of Potential and Spherical Harmonics," Toronto, University of Toronto Press, 1946.

47. J. D. Tamarkin and W. Feller: "Partial Differential Equations," lecture notes, Brown University, 1941.

48. E. C. Titchmarsh: "Theory of Functions," London, Oxford University Press, 1939.

49. E. T. Whittaker and G. N. Watson: "A Course of Modern Analysis," Cambridge, Cambridge University Press, 1952.

50. A. Zygmund: "Trigonometrical Series," New York, Chelsea, 1952.

INDEX

A CATALOG OF SELECTED
DOVER BOOKS
IN SCIENCE AND MATHEMATICS

A CATALOG OF SELECTED

DOVER BOOKS

IN SCIENCE AND MATHEMATICS

Astronomy

BURNHAM'S CELESTIAL HANDBOOK, Robert Burnham, Jr. Thorough guide to the stars beyond our solar system. Exhaustive treatment. Alphabetical by constellation: Andromeda to Cetus in Vol. 1; Chamaeleon to Orion in Vol. 2; and Pavo to Vulpecula in Vol. 3. Hundreds of illustrations. Index in Vol. 3. 2,000pp. 6⅛ x 9¼.
23567-X, 23568-8, 23673-0 Pa., Three-vol. set $46.85

THE EXTRATERRESTRIAL LIFE DEBATE, 1750–1900, Michael J. Crowe. First detailed, scholarly study in English of the many ideas that developed between 1750 and 1900 regarding the existence of intelligent extraterrestrial life. Examines ideas of Kant, Herschel, Voltaire, Percival Lowell, many other scientists and thinkers. 16 illustrations. 704pp. 5⅜ x 8½. 40675-X Pa. $19.95

A HISTORY OF ASTRONOMY, A. Pannekoek. Well-balanced, carefully reasoned study covers such topics as Ptolemaic theory, work of Copernicus, Kepler, Newton, Eddington's work on stars, much more. Illustrated. References. 521pp. 5⅜ x 8½.
65994-1 Pa. $15.95

AMATEUR ASTRONOMER'S HANDBOOK, J. B. Sidgwick. Timeless, comprehensive coverage of telescopes, mirrors, lenses, mountings, telescope drives, micrometers, spectroscopes, more. 189 illustrations. 576pp. 5⅝ x 8¼. (Available in U.S. only)
24034-7 Pa. $13.95

STARS AND RELATIVITY, Ya. B. Zel'dovich and I. D. Novikov. Vol. 1 of *Relativistic Astrophysics* by famed Russian scientists. General relativity, properties of matter under astrophysical conditions, stars and stellar systems. Deep physical insights, clear presentation. 1971 edition. References. 544pp. 5⅜ x 8½.
69424-0 Pa. $14.95

Chemistry

A SHORT HISTORY OF CHEMISTRY (3rd edition), J. R. Partington. Classic exposition explores origins of chemistry, alchemy, early medical chemistry, nature of atmosphere, theory of valency, laws and structure of atomic theory, much more. 428pp. 5⅜ x 8½. (Available in U.S. only) 65977-1 Pa. $12.95

CHEMICAL MAGIC, Leonard A. Ford. Second Edition, Revised by E. Winston Grundmeier. Over 100 unusual stunts demonstrating cold fire, dust explosions, much more. Text explains scientific principles and stresses safety precautions. 128pp. 5⅜ x 8½. 67628-5 Pa. $5.95

THE DEVELOPMENT OF MODERN CHEMISTRY, Aaron J. Ihde. Authoritative history of chemistry from ancient Greek theory to 20th-century innovation. Covers major chemists and their discoveries. 209 illustrations. 14 tables. Bibliographies. Indices. Appendices. 851pp. 5⅜ x 8½. 64235-6 Pa. $24.95

CATALYSIS IN CHEMISTRY AND ENZYMOLOGY, William P. Jencks. Exceptionally clear coverage of mechanisms for catalysis, forces in aqueous solution, carbonyl- and acyl-group reactions, practical kinetics, more. 864pp. 5⅜ x 8½.
65460-5 Pa. $19.95

THE HISTORICAL BACKGROUND OF CHEMISTRY, Henry M. Leicester. Evolution of ideas, not individual biography. Concentrates on formulation of a coherent set of chemical laws. 260pp. 5⅜ x 8½. 61053-5 Pa. $8.95

GENERAL CHEMISTRY, Linus Pauling. Revised 3rd edition of classic first-year text by Nobel laureate. Atomic and molecular structure, quantum mechanics, statistical mechanics, thermodynamics correlated with descriptive chemistry. Problems. 992pp. 5⅜ x 8½. 65622-5 Pa. $19.95

Engineering

DE RE METALLICA, Georgius Agricola. The famous Hoover translation of greatest treatise on technological chemistry, engineering, geology, mining of early modern times (1556). All 289 original woodcuts. 638pp. 6¾ x 11. 60006-8 Pa. $21.95

FUNDAMENTALS OF ASTRODYNAMICS, Roger Bate et al. Modern approach developed by U.S. Air Force Academy. Designed as a first course. Problems, exercises. Numerous illustrations. 455pp. 5⅜ x 8½. 60061-0 Pa. $12.95

DYNAMICS OF FLUIDS IN POROUS MEDIA, Jacob Bear. For advanced students of ground water hydrology, soil mechanics and physics, drainage and irrigation engineering and more. 335 illustrations. Exercises, with answers. 784pp. 6⅛ x 9¼.
65675-6 Pa. $19.95

ANALYTICAL MECHANICS OF GEARS, Earle Buckingham. Indispensable reference for modern gear manufacture covers conjugate gear-tooth action, gear-tooth profiles of various gears, many other topics. 263 figures. 102 tables. 546pp. 5⅜ x 8½.
65712-4 Pa. $16.95

ADVANCED STRENGTH OF MATERIALS, J. P. Den Hartog. Superbly written advanced text covers torsion, rotating disks, membrane stresses in shells, much more. Many problems and answers. 388pp. 5⅜ x 8½. 65407-9 Pa. $11.95

MECHANICS, J. P. Den Hartog. A classic introductory text or refresher. Hundreds of applications and design problems illuminate fundamentals of trusses, loaded beams and cables, etc. 334 answered problems. 462pp. 5⅜ x 8½. 60754-2 Pa. $12.95

MECHANICAL VIBRATIONS, J. P. Den Hartog. Classic textbook offers lucid explanations and illustrative models, applying theories of vibrations to a variety of practical industrial engineering problems. Numerous figures. 233 problems, solutions. Appendix. Index. Preface. 436pp. 5⅜ x 8½. 64785-4 Pa. $13.95

STRENGTH OF MATERIALS, J. P. Den Hartog. Full, clear treatment of basic material (tension, torsion, bending, etc.) plus advanced material on engineering methods, applications. 350 answered problems. 323pp. 5⅜ x 8½. 60755-0 Pa. $10.95

A HISTORY OF MECHANICS, René Dugas. Monumental study of mechanical principles from antiquity to quantum mechanics. Contributions of ancient Greeks, Galileo, Leonardo, Kepler, Lagrange, many others. 671pp. 5⅜ x 8½.
65632-2 Pa. $18.95

STATISTICAL MECHANICS: Principles and Applications, Terrell L. Hill. Standard text covers fundamentals of statistical mechanics, applications to fluctuation theory, imperfect gases, distribution functions, more. 448pp. 5⅜ x 8½.
65390-0 Pa. $14.95

THE VARIATIONAL PRINCIPLES OF MECHANICS, Cornelius Lanczos. Graduate level coverage of calculus of variations, equations of motion, relativistic mechanics, more. First inexpensive paperbound edition of classic treatise. Index. Bibliography. 418pp. 5⅜ x 8½.
65067-7 Pa. $14.95

THE VARIOUS AND INGENIOUS MACHINES OF AGOSTINO RAMELLI: A Classic Sixteenth-Century Illustrated Treatise on Technology, Agostino Ramelli. One of the most widely known and copied works on machinery in the 16th century. 194 detailed plates of water pumps, grain mills, cranes, more. 608pp. 9 x 12.
28180-9 Pa. $24.95

ORDINARY DIFFERENTIAL EQUATIONS AND STABILITY THEORY: An Introduction, David A. Sánchez. Brief, modern treatment. Linear equation, stability theory for autonomous and nonautonomous systems, etc. 164pp. 5⅜ x 8¼.
63828-6 Pa. $6.95

ROTARY-WING AERODYNAMICS, W. Z. Stepniewski. Clear, concise text covers aerodynamic phenomena of the rotor and offers guidelines for helicopter performance evaluation. Originally prepared for NASA. 537 figures. 640pp. 6⅛ x 9¼.
64647-5 Pa. $16.95

INTRODUCTION TO SPACE DYNAMICS, William Tyrrell Thomson. Comprehensive, classic introduction to space-flight engineering for advanced undergraduate and graduate students. Includes vector algebra, kinematics, transformation of coordinates. Bibliography. Index. 352pp. 5⅜ x 8½.
65113-4 Pa. $10.95

HISTORY OF STRENGTH OF MATERIALS, Stephen P. Timoshenko. Excellent historical survey of the strength of materials with many references to the theories of elasticity and structure. 245 figures. 452pp. 5⅜ x 8½.
61187-6 Pa. $14.95

CONSTRUCTIONS AND COMBINATORIAL PROBLEMS IN DESIGN OF EXPERIMENTS, Damaraju Raghavarao. In-depth reference work examines orthogonal Latin squares, incomplete block designs, tactical configuration, partial geometry, much more. Abundant explanations, examples. 416pp. 5⅜ x 8¼.
65685-3 Pa. $10.95

INCOMPRESSIBLE AERODYNAMICS, edited by Bryan Thwaites. Covers theoretical and experimental treatment of the uniform flow of air and viscous fluids past two-dimensional aerofoils and three-dimensional wings; many other topics. 654pp. 5⅜ x 8½.
65465-6 Pa. $16.95

Mathematics

HANDBOOK OF MATHEMATICAL FUNCTIONS WITH FORMULAS, GRAPHS, AND MATHEMATICAL TABLES, edited by Milton Abramowitz and Irene A. Stegun. Vast compendium: 29 sets of tables, some to as high as 20 places. 1,046pp. 8 x 10½. 61272-4 Pa. $29.95

CALCULUS REFRESHER FOR TECHNICAL PEOPLE, A. Albert Klaf. Covers important aspects of integral and differential calculus via 756 questions. 566 problems, most answered. 431pp. 5⅜ x 8½. 20370-0 Pa. $9.95

ASYMPTOTIC EXPANSIONS OF INTEGRALS, Norman Bleistein & Richard A. Handelsman. Best introduction to important field with applications in a variety of scientific disciplines. New preface. Problems. Diagrams. Tables. Bibliography. Index. 448pp. 5⅜ x 8½. 65082-0 Pa. $13.95

FAMOUS PROBLEMS OF GEOMETRY AND HOW TO SOLVE THEM, Benjamin Bold. Squaring the circle, trisecting the angle, duplicating the cube: learn their history, why they are impossible to solve, then solve them yourself. 128pp. 5⅜ x 8½. 24297-8 Pa. $5.95

VECTOR AND TENSOR ANALYSIS WITH APPLICATIONS, A. I. Borisenko and I. E. Tarapov. Concise introduction. Worked-out problems, solutions, exercises. 257pp. 5⅜ x 8¼. 63833-2 Pa. $9.95

THE ABSOLUTE DIFFERENTIAL CALCULUS (CALCULUS OF TENSORS), Tullio Levi-Civita. Great 20th-century mathematician's classic work on material necessary for mathematical grasp of theory of relativity. 452pp. 5⅜ x 8½.
63401-9 Pa. $11.95

AN INTRODUCTION TO ORDINARY DIFFERENTIAL EQUATIONS, Earl A. Coddington. A thorough and systematic first course in elementary differential equations for undergraduates in mathematics and science, with many exercises and problems (with answers). Index. 304pp. 5⅜ x 8½. 65942-9 Pa. $9.95

FOURIER SERIES AND ORTHOGONAL FUNCTIONS, Harry F. Davis. An incisive text combining theory and practical example to introduce Fourier series, orthogonal functions and applications of the Fourier method to boundary-value problems. 570 exercises. Answers and notes. 416pp. 5⅜ x 8½. 65973-9 Pa. $13.95

COMPUTABILITY AND UNSOLVABILITY, Martin Davis. Classic graduate-level introduction to theory of computability, usually referred to as theory of recurrent functions. New preface and appendix. 288pp. 5⅜ x 8½. 61471-9 Pa. $8.95

ASYMPTOTIC METHODS IN ANALYSIS, N. G. de Bruijn. An inexpensive, comprehensive guide to asymptotic methods—the pioneering work that teaches by explaining worked examples in detail. Index. 224pp. 5⅜ x 8½. 64221-6 Pa. $7.95

ESSAYS ON THE THEORY OF NUMBERS, Richard Dedekind. Two classic essays by great German mathematician: on the theory of irrational numbers; and on transfinite numbers and properties of natural numbers. 115pp. 5⅜ x 8½.
21010-3 Pa. $6.95

CATALOG OF DOVER BOOKS

THE GEOMETRY OF RENÉ DESCARTES, René Descartes. The great work founded analytical geometry. Original French text, Descartes's own diagrams, together with definitive Smith-Latham translation. 244pp. 5⅜ x 8½. 60068-8 Pa. $9.95

APPLIED COMPLEX VARIABLES, John W. Dettman. Step-by-step coverage of fundamentals of analytic function theory—plus lucid exposition of five important applications: Potential Theory; Ordinary Differential Equations; Fourier Transforms; Laplace Transforms; Asymptotic Expansions. 66 figures. Exercises at chapter ends. 512pp. 5⅜ x 8½. 64670-X Pa. $14.95

INTRODUCTION TO LINEAR ALGEBRA AND DIFFERENTIAL EQUATIONS, John W. Dettman. Excellent text covers complex numbers, determinants, orthonormal bases, Laplace transforms, much more. Exercises with solutions. Undergraduate level. 416pp. 5⅜ x 8½. 65191-6 Pa. $11.95

MATHEMATICAL METHODS IN PHYSICS AND ENGINEERING, John W. Dettman. Algebraically based approach to vectors, mapping, diffraction, other topics in applied math. Also generalized functions, analytic function theory, more. Exercises. 448pp. 5⅜ x 8¼. 65649-7 Pa. $12.95

THE THIRTEEN BOOKS OF EUCLID'S ELEMENTS, translated with introduction and commentary by Sir Thomas L. Heath. Definitive edition. Textual and linguistic notes, mathematical analysis. 2,500 years of critical commentary. Unabridged. 1,414pp. 5⅜ x 8½. Three-vol. set.
Vol. I: 60088-2 Pa. $10.95
Vol. II: 60089-0 Pa. $10.95
Vol. III: 60090-4 Pa. $12.95

CALCULUS OF VARIATIONS WITH APPLICATIONS, George M. Ewing. Applications-oriented introduction to variational theory develops insight and promotes understanding of specialized books, research papers. Suitable for advanced undergraduate/graduate students as primary, supplementary text. 352pp. 5⅜ x 8½. 64856-7 Pa. $9.95

COMPLEX VARIABLES, Francis J. Flanigan. Unusual approach, delaying complex algebra till harmonic functions have been analyzed from real variable viewpoint. Includes problems with answers. 364pp. 5⅜ x 8½. 61388-7 Pa. $10.95

AN INTRODUCTION TO THE CALCULUS OF VARIATIONS, Charles Fox. Graduate-level text covers variations of an integral, isoperimetrical problems, least action, special relativity, approximations, more. References. 279pp. 5⅜ x 8½. 65499-0 Pa. $8.95

CATASTROPHE THEORY FOR SCIENTISTS AND ENGINEERS, Robert Gilmore. Advanced-level treatment describes mathematics of theory grounded in the work of Poincaré, R. Thom, other mathematicians. Also important applications to problems in mathematics, physics, chemistry and engineering. 1981 edition. References. 28 tables. 397 black-and-white illustrations. xvii + 666pp. 6⅛ x 9¼. 67539-4 Pa. $17.95

INTRODUCTION TO DIFFERENCE EQUATIONS, Samuel Goldberg. Exceptionally clear exposition of important discipline with applications to sociology, psychology, economics. Many illustrative examples; over 250 problems. 260pp. 5⅜ x 8½. 65084-7 Pa. $10.95

UNBOUNDED LINEAR OPERATORS: Theory and Applications, Seymour Goldberg. Classic presents systematic treatment of the theory of unbounded linear operators in normed linear spaces with applications to differential equations. Bibliography. I99pp. 5⅜ x 8½. 64830-3 Pa. $7.95

DIFFERENTIAL GEOMETRY, Heinrich W. Guggenheimer. Local differential geometry as an application of advanced calculus and linear algebra. Curvature, transformation groups, surfaces, more. Exercises. 62 figures. 378pp. 5⅜ x 8½. 63433-7 Pa. $11.95

NUMERICAL METHODS FOR SCIENTISTS AND ENGINEERS, Richard Hamming. Classic text stresses frequency approach in coverage of algorithms, polynomial approximation, Fourier approximation, exponential approximation, other topics. Revised and enlarged 2nd edition. 721pp. 5⅜ x 8½. 65241-6 Pa. $16.95

POPULAR LECTURES ON MATHEMATICAL LOGIC, Hao Wang. Noted logician's lucid treatment of historical developments, set theory, model theory, recursion theory and constructivism, proof theory, more. 3 appendixes. Bibliography. 1981 edition. ix + 283pp. 5⅜ x 8½. 67632-3 Pa. $10.95

INTRODUCTION TO NUMERICAL ANALYSIS (2nd Edition), F. B. Hildebrand. Classic, fundamental treatment covers computation, approximation, interpolation, numerical differentiation and integration, other topics. 150 new problems. 669pp. 5⅜ x 8½. 65363-3 Pa. $16.95

THE FUNCTIONS OF MATHEMATICAL PHYSICS, Harry Hochstadt. Comprehensive treatment of orthogonal polynomials, hypergeometric functions, Hill's equation, much more. Bibliography. Index. 322pp. 5⅜ x 8½. 65214-9 Pa. $12.95

THREE PEARLS OF NUMBER THEORY, A. Y. Khinchin. Three compelling puzzles require proof of a basic law governing the world of numbers. Challenges concern van der Waerden's theorem, the Landau-Schnirelmann hypothesis and Mann's theorem, and a solution to Waring's problem. Solutions included. 64pp. 5⅜ x 8½. 40026-3 Pa. $4.95

THE PHILOSOPHY OF MATHEMATICS: An Introductory Essay, Stephan Körner. Surveys the views of Plato, Aristotle, Leibniz & Kant concerning propositions and theories of applied and pure mathematics. Introduction. Two appendices. Index. 198pp. 5⅜ x 8½. 25048-2 Pa. $8.95

INTRODUCTORY REAL ANALYSIS, A.N. Kolmogorov, S. V. Fomin. Translated by Richard A. Silverman. Self-contained, evenly paced introduction to real and functional analysis. Some 350 problems. 403pp. 5⅜ x 8½. 61226-0 Pa. $12.95

APPLIED ANALYSIS, Cornelius Lanczos. Classic work on analysis and design of finite processes for approximating solution of analytical problems. Algebraic equations, matrices, harmonic analysis, quadrature methods, much more. 559pp. 5⅜ x 8½. 65656-X Pa. $16.95

AN INTRODUCTION TO ALGEBRAIC STRUCTURES, Joseph Landin. Superb self-contained text covers "abstract algebra": sets and numbers, theory of groups, theory of rings, much more. Numerous well-chosen examples, exercises. 247pp. 5⅜ x 8½. 65940-2 Pa. $10.95

SPECIAL FUNCTIONS, N. N. Lebedev. Translated by Richard Silverman. Famous Russian work treating more important special functions, with applications to specific problems of physics and engineering. 38 figures. 308pp. 5⅜ x 8½. 60624-4 Pa. $9.95

QUALITATIVE THEORY OF DIFFERENTIAL EQUATIONS, V. V. Nemytskii and V.V. Stepanov. Classic graduate-level text by two prominent Soviet mathematicians covers classical differential equations as well as topological dynamics and ergodic theory. Bibliographies. 523pp. 5⅜ x 8½. 65954-2 Pa. $14.95

NUMBER THEORY AND ITS HISTORY, Oystein Ore. Unusually clear, accessible introduction covers counting, properties of numbers, prime numbers, much more. Bibliography. 380pp. 5⅜ x 8½. 65620-9 Pa. $10.95

THEORY OF MATRICES, Sam Perlis. Outstanding text covering rank, nonsingularity and inverses in connection with the development of canonical matrices under the relation of equivalence, and without the intervention of determinants. Includes exercises. 237pp. 5⅜ x 8½. 66810-X Pa. $8.95

OPTIMIZATION THEORY WITH APPLICATIONS, Donald A. Pierre. Broad spectrum approach to important topic. Classical theory of minima and maxima, calculus of variations, simplex technique and linear programming, more. Many problems, examples. 640pp. 5⅜ x 8½. 65205-X Pa. $17.95

INTRODUCTION TO ANALYSIS, Maxwell Rosenlicht. Unusually clear, accessible coverage of set theory, real number system, metric spaces, continuous functions, Riemann integration, multiple integrals, more. Wide range of problems. Undergraduate level. Bibliography. 254pp. 5⅜ x 8½. 65038-3 Pa. $9.95

MODERN NONLINEAR EQUATIONS, Thomas L. Saaty. Emphasizes practical solution of problems; covers seven types of equations. ". . . a welcome contribution to the existing literature...."–*Math Reviews*. 490pp. 5⅜ x 8½. 64232-1 Pa. $13.95

MATRICES AND LINEAR ALGEBRA, Hans Schneider and George Phillip Barker. Basic textbook covers theory of matrices and its applications to systems of linear equations and related topics such as determinants, eigenvalues and differential equations. Numerous exercises. 432pp. 5⅜ x 8½. 66014-1 Pa. $12.95

GEOMETRY OF COMPLEX NUMBERS, Hans Schwerdtfeger. Illuminating, widely praised book on analytic geometry of circles, the Moebius transformation, and two-dimensional non-Euclidean geometries. 200pp. 5⅝ x 8¼. 63830-8 Pa. $8.95

MATHEMATICS APPLIED TO CONTINUUM MECHANICS, Lee A. Segel. Analyzes models of fluid flow and solid deformation. For upper-level math, science and engineering students. 608pp. 5⅜ x 8½. 65369-2 Pa. $14.95

ELEMENTS OF REAL ANALYSIS, David A. Sprecher. Classic text covers fundamental concepts, real number system, point sets, functions of a real variable, Fourier series, much more. Over 500 exercises. 352pp. 5⅜ x 8½. 65385-4 Pa. $11.95

AN INTRODUCTION TO MATRICES, SETS AND GROUPS FOR SCIENCE STUDENTS, G. Stephenson. Concise, readable text introduces sets, groups, and most importantly, matrices to undergraduate students of physics, chemistry, and engineering. Problems. 164pp. 5⅜ x 8½. 65077-4 Pa. $7.95

SET THEORY AND LOGIC, Robert R. Stoll. Lucid introduction to unified theory of mathematical concepts. Set theory and logic seen as tools for conceptual understanding of real number system. 496pp. 5⅜ x 8¼.   63829-4 Pa. $14.95

LECTURES ON CLASSICAL DIFFERENTIAL GEOMETRY, Second Edition, Dirk J. Struik. Excellent brief introduction covers curves, theory of surfaces, fundamental equations, geometry on a surface, conformal mapping, other topics. Problems. 240pp. 5⅜ x 8½.   65609-8 Pa. $9.95

ORDINARY DIFFERENTIAL EQUATIONS, Morris Tenenbaum and Harry Pollard. Exhaustive survey of ordinary differential equations for undergraduates in mathematics, engineering, science. Thorough analysis of theorems. Diagrams. Bibliography. Index. 818pp. 5⅜ x 8½.   64940-7 Pa. $19.95

INTEGRAL EQUATIONS, F. G. Tricomi. Authoritative, well-written treatment of extremely useful mathematical tool with wide applications. Volterra Equations, Fredholm Equations, much more. Advanced undergraduate to graduate level. Exercises. Bibliography. 238pp. 5⅜ x 8½.   64828-1 Pa. $8.95

FOURIER SERIES, Georgi P. Tolstov. Translated by Richard A. Silverman. A valuable addition to the literature on the subject, moving clearly from subject to subject and theorem to theorem. 107 problems, answers. 336pp. 5⅜ x 8½. 63317-9 Pa. $11.95

DISTRIBUTION THEORY AND TRANSFORM ANALYSIS: An Introduction to Generalized Functions, with Applications, A. H. Zemanian. Provides basics of distribution theory, describes generalized Fourier and Laplace transformations. Numerous problems. 384pp. 5⅜ x 8½.   65479-6 Pa. $13.95

TENSOR CALCULUS, J.L. Synge and A. Schild. Widely used introductory text covers spaces and tensors, basic operations in Riemannian space, non-Riemannian spaces, etc. 324pp. 5⅜ x 8¼.   63612-7 Pa. $11.95

CALCULUS OF VARIATIONS, Robert Weinstock. Basic introduction covering isoperimetric problems, theory of elasticity, quantum mechanics, electrostatics, etc. Exercises throughout. 326pp. 5⅜ x 8½.   63069-2 Pa. $9.95

THE CONTINUUM: A Critical Examination of the Foundation of Analysis, Hermann Weyl. Classic of 20th-century foundational research deals with the conceptual problem posed by the continuum. 156pp. 5⅜ x 8½.   67982-9 Pa. $8.95

CHALLENGING MATHEMATICAL PROBLEMS WITH ELEMENTARY SOLUTIONS, A. M. Yaglom and I. M. Yaglom. Over 170 challenging problems on probability theory, combinatorial analysis, points and lines, topology, convex polygons, many other topics. Solutions. Total of 445pp. 5⅜ x 8½. Two-vol. set.
Vol. I: 65536-9 Pa. $8.95
Vol. II: 65537-7 Pa. $7.95

A SURVEY OF NUMERICAL MATHEMATICS, David M. Young and Robert Todd Gregory. Broad self-contained coverage of computer-oriented numerical algorithms for solving various types of mathematical problems in linear algebra, ordinary and partial, differential equations, much more. Exercises. Total of 1,248pp. 5⅜ x 8½. Two volumes.
Vol. I: 65691-8 Pa. $16.95
Vol. II: 65692-6 Pa. $16.95

INTRODUCTION TO PARTIAL DIFFERENTIAL EQUATIONS WITH APPLICATIONS, E. C. Zachmanoglou and Dale W. Thoe. Essentials of partial differential equations applied to common problems in engineering and the physical sciences. Problems and answers. 416pp. 5⅜ x 8½.								65251-3 Pa. $11.95

THE THEORY OF GROUPS, Hans J. Zassenhaus. Well-written graduate-level text acquaints reader with group-theoretic methods and demonstrates their usefulness in mathematics. Axioms, the calculus of complexes, homomorphic mapping, p-group theory, more. Many proofs shorter and more transparent than older ones. 276pp. 5⅜ x 8½.								40922-8 Pa. $12.95

GENERALIZED INTEGRAL TRANSFORMATIONS, A.H. Zemanian. Graduate-level study of recent generalizations of the Laplace, Mellin, Hankel, K. Weierstrass, convolution and other simple transformations. Bibliography. 320pp. 5⅜ x 8½.
								65375-7 Pa. $8.95

Math–Decision Theory, Statistics, Probability

ELEMENTARY DECISION THEORY, Herman Chernoff and Lincoln E. Moses. Clear introduction to statistics and statistical theory covers data processing, probability and random variables, testing hypotheses, much more. Exercises. 364pp. 5⅜ x 8½.								65218-1 Pa. $12.95

STATISTICS MANUAL, Edwin L. Crow et al. Comprehensive, practical collection of classical and modern methods prepared by U.S. Naval Ordnance Test Station. Stress on use. Basics of statistics assumed. 288pp. 5⅜ x 8½.			60599-X Pa. $8.95

SOME THEORY OF SAMPLING, William Edwards Deming. Analysis of the problems, theory and design of sampling techniques for social scientists, industrial managers and others who find statistics increasingly important in their work. 61 tables. 90 figures. xvii + 602pp. 5⅜ x 8½.				64684-X Pa. $16.95

STATISTICAL ADJUSTMENT OF DATA, W. Edwards Deming. Introduction to basic concepts of statistics, curve fitting, least squares solution, conditions without parameter, conditions containing parameters. 26 exercises worked out. 271pp. 5⅜ x 8½.
								64685-8 Pa. $9.95

LINEAR PROGRAMMING AND ECONOMIC ANALYSIS, Robert Dorfman, Paul A. Samuelson and Robert M. Solow. First comprehensive treatment of linear programming in standard economic analysis. Game theory, modern welfare economics, Leontief input-output, more. 525pp. 5⅜ x 8½.			65491-5 Pa. $17.95

DICTIONARY/OUTLINE OF BASIC STATISTICS, John E. Freund and Frank J. Williams. A clear concise dictionary of over 1,000 statistical terms and an outline of statistical formulas covering probability, nonparametric tests, much more. 208pp. 5⅜ x 8½.								66796-0 Pa.$8.95

PROBABILITY: An Introduction, Samuel Goldberg. Excellent basic text covers set theory, probability theory for finite sample spaces, binomial theorem, much more. 360 problems. Bibliographies. 322pp. 5⅜ x 8½.				65252-1 Pa. $10.95

GAMES AND DECISIONS: Introduction and Critical Survey, R. Duncan Luce and Howard Raiffa. Superb nontechnical introduction to game theory, primarily applied to social sciences. Utility theory, zero-sum games, n-person games, decision-making, much more. Bibliography. 509pp. 5⅜ x 8½.   65943-7 Pa. $14.95

FIFTY CHALLENGING PROBLEMS IN PROBABILITY WITH SOLUTIONS, Frederick Mosteller. Remarkable puzzlers, graded in difficulty, illustrate elementary and advanced aspects of probability. Detailed solutions. 88pp. 5⅜ x 8½.
65355-2 Pa. $4.95

PROBABILITY THEORY: A Concise Course, Y. A. Rozanov. Highly readable, self-contained introduction covers combination of events, dependent events, Bernoulli trials, etc. Translation by Richard Silverman. 148pp. 5⅜ x 8¼.
63544-9 Pa. $8.95

STATISTICAL METHOD FROM THE VIEWPOINT OF QUALITY CONTROL, Walter A. Shewhart. Important text explains regulation of variables, uses of statistical control to achieve quality control in industry, agriculture, other areas. 192pp. 5⅜ x 8½.   65232-7 Pa. $8.95

THE COMPLEAT STRATEGYST: Being a Primer on the Theory of Games of Strategy, J. D. Williams. Highly entertaining classic describes, with many illustrated examples, how to select best strategies in conflict situations. Prefaces. Appendices. 268pp. 5⅜ x 8½.   25101-2 Pa. $8.95

Math–History of

A SHORT ACCOUNT OF THE HISTORY OF MATHEMATICS, W. W. Rouse Ball. One of clearest, most authoritative surveys from the Egyptians and Phoenicians through 19th-century figures such as Grassman, Galois, Riemann. Fourth edition. 522pp. 5⅜ x 8½.   20630-0 Pa. $13.95

THE HISTORICAL ROOTS OF ELEMENTARY MATHEMATICS, Lucas N. H. Bunt, Phillip S. Jones, and Jack D. Bedient. Fundamental underpinnings of modern arithmetic, algebra, geometry and number systems derived from ancient civilizations. 320pp. 5⅜ x 8½.   25563-8 Pa. $9.95

GAMES, GODS & GAMBLING: A History of Probability and Statistical Ideas, F. N. David. Episodes from the lives of Galileo, Fermat, Pascal, and others illustrate this fascinating account of the roots of mathematics. Features thought-provoking references to classics, archaeology, biography, poetry. 1962 edition. 304pp. 5⅜ x 8½. (USO)   40023-9 Pa. $9.95

HISTORY OF MATHEMATICS, David E. Smith. Nontechnical survey from ancient Greece and Orient to late 19th century; evolution of arithmetic, geometry, trigonometry, calculating devices, algebra, the calculus. 362 illustrations. 1,355pp. 5⅜ x 8½. Two-vol. set.   Vol. I: 20429-4 Pa. $13.95
Vol. II: 20430-8 Pa. $14.95

A CONCISE HISTORY OF MATHEMATICS, Dirk J. Struik. The best brief history of mathematics. Stresses origins and covers every major figure from ancient Near East to 19th century. 41 illustrations. 195pp. 5⅜ x 8½.   60255-9 Pa. $8.95

THE HISTORY OF THE CALCULUS AND ITS CONCEPTUAL DEVELOP-MENT, Carl B. Boyer. Origins in antiquity, medieval contributions, work of Newton, Leibniz, rigorous formulation. Treatment is verbal. 346pp. 5⅜ x 8½. 60509-4 Pa. $9.95

Math–Topology

ELEMENTARY CONCEPTS OF TOPOLOGY, Paul Alexandroff. Elegant, intuitive approach to topology from set-theoretic topology to Betti groups; how concepts of topology are useful in math and physics. 25 figures. 57pp. 5⅜ x 8½.
60747-X Pa. $4.95

COMBINATORIAL TOPOLOGY, P. S. Alexandrov. Clearly written, well-organized, three-part text begins by dealing with certain classic problems without using the formal techniques of homology theory and advances to the central concept, the Betti groups. Numerous detailed examples. 654pp. 5⅜ x 8½.   40179-0 Pa. $18.95

EXPERIMENTS IN TOPOLOGY, Stephen Barr. Classic, lively explanation of one of the byways of mathematics. Klein bottles, Moebius strips, projective planes, map coloring, problem of the Koenigsberg bridges, much more, described with clarity and wit. 43 figures. 210pp. 5⅜ x 8½.   25933-1 Pa. $8.95

CONFORMAL MAPPING ON RIEMANN SURFACES, Harvey Cohn. Lucid, insightful book presents ideal coverage of subject. 334 exercises make book perfect for self-study. 55 figures. 352pp. 5⅜ x 8¼.   64025-6 Pa. $11.95

CURVATURE AND HOMOLOGY: Enlarged Edition, Samuel I. Goldberg. Revised edition examines topology of differentiable manifolds; curvature, homology of Riemannian manifolds; compact Lie groups; complex manifolds; curvature, homology of Kaehler manifolds. New Preface. Four new appendixes. 416pp. 5⅜ x 8½.
40207-X Pa. $14.95

TOPOLOGY, John G. Hocking and Gail S. Young. Superb one-year course in classical topology. Topological spaces and functions, point-set topology, much more. Examples and problems. Bibliography. Index. 384pp. 5⅜ x 8¼.   65676-4 Pa. $11.95

THE FOUR-COLOR PROBLEM: Assaults and Conquest, Thomas L. Saaty and Paul G. Kainen. Engrossing, comprehensive account of the century-old combinatorial topological problem, its history and solution. Bibliographies. Index. 110 figures. 228pp. 5⅜ x 8½.   65092-8 Pa. $7.95

Meteorology

PRINCIPLES OF METEOROLOGICAL ANALYSIS, Walter J. Saucier. Highly respected, abundantly illustrated classic reviews atmospheric variables, hydrostatics, static stability, various analyses (scalar, cross-section, isobaric, isentropic, more). For intermediate meteorology students. 454pp. 6⅛ x 9¼.   65979-8 Pa. $14.95

LIGHTNING, Martin A. Uman. Revised, updated edition of classic work on the physics of lightning. Phenomena, terminology, measurement, photography, spectroscopy, thunder, more. Reviews recent research. Bibliography. Indices. 320pp. 5⅜ x 8¼.   64575-4 Pa. $8.95

Physics

OPTICAL RESONANCE AND TWO-LEVEL ATOMS, L. Allen and J. H. Eberly. Clear, comprehensive introduction to basic principles behind all quantum optical resonance phenomena. 53 illustrations. Preface. Index. 256pp. 5⅜ x 8½.
65533-4 Pa. $10.95

ULTRASONIC ABSORPTION: An Introduction to the Theory of Sound Absorption and Dispersion in Gases, Liquids and Solids, A. B. Bhatia. Standard reference in the field provides a clear, systematically organized introductory review of fundamental concepts for advanced graduate students, research workers. Numerous diagrams. Bibliography. 440pp. 5⅜ x 8½.
64917-2 Pa. $11.95

QUANTUM THEORY, David Bohm. This advanced undergraduate-level text presents the quantum theory in terms of qualitative and imaginative concepts, followed by specific applications worked out in mathematical detail. Preface. Index. 655pp. 5⅜ x 8½.
65969-0 Pa. $15.95

ATOMIC PHYSICS (8th edition), Max Born. Nobel laureate's lucid treatment of kinetic theory of gases, elementary particles, nuclear atom, wave-corpuscles, atomic structure and spectral lines, much more. Over 40 appendices, bibliography. 495pp. 5⅜ x 8½.
65984-4 Pa. $13.95

AN INTRODUCTION TO HAMILTONIAN OPTICS, H. A. Buchdahl. Detailed account of the Hamiltonian treatment of aberration theory in geometrical optics. Many classes of optical systems defined in terms of the symmetries they possess. Problems with detailed solutions. 1970 edition. xv + 360pp. 5⅜ x 8½.
67597-1 Pa. $10.95

HYDRODYNAMIC AND HYDROMAGNETIC STABILITY, S. Chandrasekhar. Lucid examination of the Rayleigh-Benard problem; clear coverage of the theory of instabilities causing convection. 704pp. 5⅜ x 8¼.
64071-X Pa. $17.95

INVESTIGATIONS ON THE THEORY OF THE BROWNIAN MOVEMENT, Albert Einstein. Five papers (1905–8) investigating dynamics of Brownian motion and evolving elementary theory. Notes by R. Fürth. 122pp. 5⅜ x 8½.
60304-0 Pa. $5.95

THE PHYSICS OF WAVES, William C. Elmore and Mark A. Heald. Unique overview of classical wave theory. Acoustics, optics, electromagnetic radiation, more. Ideal as classroom text or for self-study. Problems. 477pp. 5⅜ x 8½.
64926-1 Pa. $14.95

THIRTY YEARS THAT SHOOK PHYSICS: The Story of Quantum Theory, George Gamow. Lucid, accessible introduction to influential theory of energy and matter. Careful explanations of Dirac's anti-particles, Bohr's model of the atom, much more. 12 plates. Numerous drawings. 240pp. 5⅜ x 8½.
24895-X Pa. $7.95

ELECTRONIC STRUCTURE AND THE PROPERTIES OF SOLIDS: The Physics of the Chemical Bond, Walter A. Harrison. Innovative text offers basic understanding of the electronic structure of covalent and ionic solids, simple metals, transition metals and their compounds. Problems. 1980 edition. 582pp. 6⅛ x 9¼.
66021-4 Pa. $19.95

PHYSICAL PRINCIPLES OF THE QUANTUM THEORY, Werner Heisenberg. Nobel Laureate discusses quantum theory, uncertainty, wave mechanics, work of Dirac, Schroedinger, Compton, Wilson, Einstein, etc. 184pp. 5⅜ x 8½.
60113-7 Pa. $8.95

ATOMIC SPECTRA AND ATOMIC STRUCTURE, Gerhard Herzberg. One of best introductions; especially for specialist in other fields. Treatment is physical rather than mathematical. 80 illustrations. 257pp. 5⅜ x 8½.   60115-3 Pa. $7.95

AN INTRODUCTION TO STATISTICAL THERMODYNAMICS, Terrell L. Hill. Excellent basic text offers wide-ranging coverage of quantum statistical mechanics, systems of interacting molecules, quantum statistics, more. 523pp. 5⅜ x 8½.
65242-4 Pa. $13.95

THEORETICAL PHYSICS, Georg Joos, with Ira M. Freeman. Classic overview covers essential math, mechanics, electromagnetic theory, thermodynamics, quantum mechanics, nuclear physics, other topics. First paperback edition. xxiii + 885pp. 5⅜ x 8½.   65227-0 Pa. $21.95

BOUNDARY VALUE PROBLEMS OF HEAT CONDUCTION, M. Necati Özisik. Systematic, comprehensive treatment of modern mathematical methods of solving problems in heat conduction and diffusion. Numerous examples and problems. Selected references. Appendices. 505pp. 5⅜ x 8½.   65990-9 Pa. $12.95

PROBLEMS AND SOLUTIONS IN QUANTUM CHEMISTRY AND PHYSICS, Charles S. Johnson, Jr. and Lee G. Pedersen. Unusually varied problems, detailed solutions in coverage of quantum mechanics, wave mechanics, angular momentum, molecular spectroscopy, scattering theory, more. 280 problems plus 139 supplementary exercises. 430pp. 6½ x 9¼.   65236-X Pa. $14.95

THEORETICAL SOLID STATE PHYSICS, Vol. 1: Perfect Lattices in Equilibrium; Vol. II: Non-Equilibrium and Disorder, William Jones and Norman H. March. Monumental reference work covers fundamental theory of equilibrium properties of perfect crystalline solids, non-equilibrium properties, defects and disordered systems. Appendices. Problems. Preface. Diagrams. Index. Bibliography. Total of 1,301pp. 5⅜ x 8½. Two volumes.   Vol. I: 65015-4 Pa. $16.95
Vol. II: 65016-2 Pa. $16.95

A TREATISE ON ELECTRICITY AND MAGNETISM, James Clerk Maxwell. Important foundation work of modern physics. Brings to final form Maxwell's theory of electromagnetism and rigorously derives his general equations of field theory. 1,084pp. 5⅜ x 8½. Two-vol. set.   Vol. I: 60636-8 Pa. $14.95
Vol. II: 60637-6 Pa. $12.95

OPTICKS, Sir Isaac Newton. Newton's own experiments with spectroscopy, colors, lenses, reflection, refraction, etc., in language the layman can follow. Foreword by Albert Einstein. 532pp. 5⅜ x 8½.   60205-2 Pa. $13.95

THEORY OF ELECTROMAGNETIC WAVE PROPAGATION, Charles Herach Papas. Graduate-level study discusses the Maxwell field equations, radiation from wire antennas, the Doppler effect and more. xiii + 244pp. 5⅜ x 8½. 65678-0 Pa. $9.95

INTRODUCTION TO QUANTUM MECHANICS With Applications to Chemistry, Linus Pauling & E. Bright Wilson, Jr. Classic undergraduate text by Nobel Prize winner applies quantum mechanics to chemical and physical problems. Numerous tables and figures enhance the text. Chapter bibliographies. Appendices. Index. 468pp. 5⅜ x 8½. 64871-0 Pa. $12.95

METHODS OF THERMODYNAMICS, Howard Reiss. Outstanding text focuses on physical technique of thermodynamics, typical problem areas of understanding, and significance and use of thermodynamic potential. 1965 edition. 238pp. 5⅜ x 8½. 69445-3 Pa. $8.95

TENSOR ANALYSIS FOR PHYSICISTS, J. A. Schouten. Concise exposition of the mathematical basis of tensor analysis, integrated with well-chosen physical examples of the theory. Exercises. Index. Bibliography. 289pp. 5⅜ x 8½. 65582-2 Pa. $10.95

RELATIVITY IN ILLUSTRATIONS, Jacob T. Schwartz. Clear nontechnical treatment makes relativity more accessible than ever before. Over 60 drawings illustrate concepts more clearly than text alone. Only high school geometry needed. Bibliography. 128pp. 6⅛ x 9¼. 25965-X Pa. $7.95

THE ELECTROMAGNETIC FIELD, Albert Shadowitz. Comprehensive undergraduate text covers basics of electric and magnetic fields, builds up to electromagnetic theory. Also related topics, including relativity. Over 900 problems. 768pp. 5⅝ x 8¼. 65660-8 Pa. $19.95

GREAT EXPERIMENTS IN PHYSICS: Firsthand Accounts from Galileo to Einstein, edited by Morris H. Shamos. 25 crucial discoveries: Newton's laws of motion, Chadwick's study of the neutron, Hertz on electromagnetic waves, more. Original accounts clearly annotated. 370pp. 5⅜ x 8½. 25346-5 Pa. $11.95

RELATIVITY, THERMODYNAMICS AND COSMOLOGY, Richard C. Tolman. Landmark study extends thermodynamics to special, general relativity; also applications of relativistic mechanics, thermodynamics to cosmological models. 501pp. 5⅜ x 8½. 65383-8 Pa. $15.95

LIGHT SCATTERING BY SMALL PARTICLES, H. C. van de Hulst. Comprehensive treatment including full range of useful approximation methods for researchers in chemistry, meteorology and astronomy. 44 illustrations. 470pp. 5⅜ x 8½. 64228-3 Pa. $12.95

STATISTICAL PHYSICS, Gregory H. Wannier. Classic text combines thermodynamics, statistical mechanics and kinetic theory in one unified presentation of thermal physics. Problems with solutions. Bibliography. 532pp. 5⅜ x 8½. 65401-X Pa. $14.95

Prices subject to change without notice.

Available at your book dealer or write for free Dover Mathematics and Science Catalog (59065-8) to Dept. GI, Dover Publications, Inc., 31 East 2nd St., Mineola, N.Y. 11501. Dover publishes more than 250 books each year on science, elementary and advanced mathematics, biology, music, art, literature, history, social sciences, and other subjects.